FORCING FOR
MATHEMATICIANS

FORCING FOR MATHEMATICIANS

Nik Weaver
Washington University in St. Louis, USA

World Scientific

NEW JERSEY · LONDON · SINGAPORE · BEIJING · SHANGHAI · HONG KONG · TAIPEI · CHENNAI

Published by

World Scientific Publishing Co. Pte. Ltd.

5 Toh Tuck Link, Singapore 596224

USA office: 27 Warren Street, Suite 401-402, Hackensack, NJ 07601

UK office: 57 Shelton Street, Covent Garden, London WC2H 9HE

Library of Congress Cataloging-in-Publication Data
Weaver, Nik, author.
    Forcing for mathematicians / by Nik Weaver (Washington University in St. Louis, USA).
      pages cm
    Includes bibliographical references and index.
    ISBN 978-9814566001 (hardcover : alk. paper)
    1. Forcing (Model theory) 2. Set theory. 3. Axiom of choice. 4. Continuum hypothesis.
I. Title.
    QA9.7.W435 2014
    511.3'4--dc23
                                    2013047943

British Library Cataloguing-in-Publication Data
A catalogue record for this book is available from the British Library.

Printed in Singapore by World Scientific Printers.

We're doing set theory, so "sets" are sets of sets.

D. A. Martin

Preface

This book was written for mathematicians who want to learn the basic machinery of forcing. No background in logic is assumed, beyond the facility with formal syntax which should be second nature to any well-trained mathematician.

As a student I found this subject confusing, so I have tried to explain it in a way that I imagine would have made it easier for me to grasp. At a technical level forcing is no more complex than any other serious mathematical topic; the difficulty is more a matter of needing to pay attention to subtle distinctions which might seem pedantic but are actually very important, such as the distinction between an axiom and an axiom scheme, or between reasoning in the target theory and reasoning in the metatheory. In other expositions this last point is often clarified only after the basic theory has been developed, almost as an afterthought. I felt that some mystification could be avoided by getting metatheoretic issues off the table at the start.

I have altered some standard definitions (most profitably, the definition of a P-name) in order to simplify the presentation. You're welcome.

A unique feature of the book is its emphasis on applications outside of set theory which were previously only available in the primary literature.

Independence results raise deep philosophical questions about the nature of mathematics. I have strong views on this subject and I felt it would be disingenuous not to mention them; however, I have confined my remarks on this topic to Chapter 30, so they can easily be ignored by readers who are not interested in such discussions.

This work was partially supported by NSF grant DMS-1067726.

Nik Weaver

Contents

Chapter 1

Peano Arithmetic

Forcing is a powerful technique for proving consistency and independence results in relation to axiomatic set theory. A statement is *consistent* with a given family of axioms if it cannot be disproven on the basis of those axioms, and *independent* of them if it can be neither proven nor disproven. When we have established that some assertion is consistent, there is still hope that it might actually be provable from the axioms, but once we have shown it is independent the matter is closed.

The preeminent historical example of an independent statement is Euclid's fifth axiom, the parallel postulate:

(5) Given a line and a point not on that line, at most one line can be drawn through the given point that is parallel to the given line.

This statement is independent of the other four axioms of Euclidean geometry:

(1) A straight line can be drawn between any two points.
(2) A line segment can be extended indefinitely in both directions.
(3) A circle can be drawn with any center and any radius.
(4) All right angles are equal to each other.

How do we know this? The parallel postulate is consistent with the other four axioms because all five statements hold in the standard Euclidean plane. Its negation is consistent with the other four axioms because the first four axioms hold in a hyperbolic plane (taking "lines" to be geodesics), but the parallel postulate fails. Thus, we may assume either the parallel postulate or its negation without fear of contradiction.

No doubt the reader is already familiar to some degree with this example and is not about to raise any objections to the conclusion we just reached.

But there is room for criticism. The problem is that Euclid's "axioms" are stated in an informal manner that apparently presupposes an intuitive grasp of the flat plane they are intended to describe. In fact they hardly qualify as axioms in the modern sense. To be fair, Euclid does preface his axioms with informal "definitions" of the terms appearing in them, but several of these are also quite vague (e.g., "a line is a breadthless length") and again presuppose some implicit knowledge of the subject matter.

Thus, there is a legitimate question as to whether it is completely clear that this implicit knowledge assumed by Euclid is compatible with the hyperbolic plane example. I do not wish to argue this point, only to emphasize the desirability of setting up a purely formal axiomatic system equipped with a precise symbolic language and well-defined rules of inference. If the system is ambiguous in any way then we cannot consider consistency and independence to be rigorous mathematical concepts.

Peano arithmetic, usually abbreviated PA, is a good example of a formal axiomatic system. It is simple enough to be described in detail. These details don't matter so much for us, but seeing them once may help give the reader a clearer sense of the way axiomatic systems work.

The language of PA is specified as follows. We start with an infinite list of variables $x, y, \ldots$; a constant symbol 0; symbols for the addition, multiplication, and successor operations $(+, \cdot, ')$; and parentheses. The variables are to be thought of as ranging over the natural numbers, and the successor symbol as representing the operation of adding 1. A *term* is any grammatical expression built up from these components, e.g., something like $0'' + x \cdot y'$ (with parentheses omitted here for the sake of readability), and an *atomic formula* is a statement of the form $t_1 = t_2$ where t_1 and t_2 are terms. Finally, a *formula* is any statement built up from atomic formulas using parentheses and the logical symbols $\neg$ (not), $\rightarrow$ (implies), and $\forall$ (for all).

For the sake of economy, we can limit ourselves to these three logical symbols and regard expressions involving the symbols $\vee$ (or), $\wedge$ (and), $\leftrightarrow$ (if and only if), and $\exists$ (there exists) as abbreviating longer expressions involving only $\neg$, $\rightarrow$, and $\forall$. For instance, $\phi \vee \psi$ is equivalent to $\neg\phi \rightarrow \psi$. Thus statements like "$x$ is prime" can be rendered symbolically, say as
$$\neg(x = 0') \wedge (\forall y)(\forall z)[x = y \cdot z \rightarrow (y = 0' \vee z = 0')],$$
and then translated into a form that uses only $\neg$, $\rightarrow$, and $\forall$. Evidently, this simple language is flexible enough to express a large variety of elementary number-theoretic assertions: every number is a sum of four squares, there is a prime pair greater than any number, etc.

The axioms of PA come in three groups. First, we have *logical axioms* which represent general logical truths, starting with

L1 $\phi \to (\psi \to \phi)$
L2 $[\phi \to (\psi \to \theta)] \to [(\phi \to \psi) \to (\phi \to \theta)]$
L3 $(\neg\psi \to \neg\phi) \to (\phi \to \psi)$.

Properly speaking, the preceding are not axioms but axiom schemes, meaning that they are to be thought of as templates which can be used to generate infinitely many axioms by replacing ϕ, ψ, and θ with any formulas. Also falling under the rubric of "logical axioms" are two schemes pertaining to quantification,

L4 $(\forall x)(\phi \to \psi) \to (\phi \to (\forall x)\psi)$,

where ϕ and ψ are any formulas such that ϕ contains no unquantified appearance of x, and

L5 $(\forall x)\phi(x) \to \phi(t)$,

where ϕ is any formula and t is any term that can be substituted for x in $\phi(x)$ without any of its variables becoming quantified. (This restriction on t prevents disasters like $(\forall x)(\exists y)(y = x) \to (\exists y)(y = y')$.)

Next, we have *equality axioms* which describe basic properties of equality. The axioms

E1 $x = x$
E2 $x = y \to y = x$
E3 $x = y \to (y = z \to x = z)$
E4 $x = y \to x' = y'$

suffice here. Axioms analogous to E4 for the other operations $+$ and $\cdot$ need not be added separately because they can be proven, within PA, from E4.

Finally, PA includes the *non-logical axioms*

(1) $\neg(0 = x')$
(2) $x' = y' \to x = y$
(3) $x + 0 = x$
(4) $x + y' = (x + y)'$
(5) $x \cdot 0 = 0$
(6) $x \cdot y' = x \cdot y + x$
(7) $\phi(0) \to [(\forall x)(\phi(x) \to \phi(x')) \to (\forall x)\phi(x)]$

which represent truths specific to arithmetic. These are all single axioms except for the induction scheme (7), in which ϕ can be any formula. The formal system is completed by specifying three *rules of inference*:

I1 from ϕ and $\phi \rightarrow \psi$, infer ψ
I2 from ϕ, infer $(\forall x)\phi$
I3 from ϕ, infer $\tilde{\phi}$

which are also presented schematically; in the last scheme $\tilde{\phi}$ can be any formula obtained from ϕ by renaming variables. (This last scheme is necessary because earlier axioms were stated in terms of the particular variables x, y, and z.) A *theorem* of PA is any formula which is either an axiom or can be derived from finitely many axioms by the application of finitely many rules of inference.

Peano arithmetic attains a perfect degree of precision. It would not be terribly difficult to write a computer program that would mechanically print out all the theorems provable in PA. The question whether some formula in the language of arithmetic is or is not a theorem is completely precise.

Although the Peano axioms are simple, they are quite powerful. Loosely speaking, all "ordinary" reasoning in elementary number theory can be formalized in PA. Indeed, we might define *elementary number theory* to be that part of number theory which can be formally codified in PA.

There are arithmetical assertions which are known to be independent of PA, although it is not so easy to come up with examples. The existence of such assertions goes back to Gödel, though we now know of simpler examples than the ones he constructed. But this topic falls outside our purview.

The details of how one actually formalizes reasoning in PA need not concern us here. It may be a valuable skill to be able to convert human-readable proofs into the machine language of a formal system, but developing that skill is not our purpose. So we will simply ask the reader to accept that elementary reasoning about numbers can be carried out in PA. More information on formalization of proofs can be found in any good introduction to mathematical logic.

Chapter 2

Zermelo-Fraenkel Set Theory

The standard formal system for reasoning about sets is ZFC, Zermelo-Fraenkel set theory with the axiom of choice. This is a *pure* set theory, meaning that there are no objects besides sets. Every element of a set is another set.

The formal language of set theory is even simpler than the language of arithmetic. We have an infinite list of variables $x, y, \ldots$, and the only atomic formulas are those of the form $x \in y$ or $x = y$, with any variables in place of x and y. There are no "terms" other than individual variables. As in arithmetic, arbitrary formulas are built up from atomic formulas using $\neg$, $\rightarrow$, and $\forall$, and we justify the informal use of other logical symbols by reducing them to these symbols. The logical axiom schemes of ZFC are the same schemes L1-L5 used in PA, although the actual axioms generated by these schemes are different, because the pool of formulas available to be substituted into the templates is different. We need two equality axioms, $x = y \rightarrow (x \in z \rightarrow y \in z)$ and $x = y \rightarrow (z \in x \rightarrow z \in y)$. The rules of inference for ZFC are generated by the same schemes I1-I3 used in PA.

The non-logical axioms of ZFC consist of seven individual axioms and two schemes. We will state them informally, but it would not be hard to translate them into the formal language of set theory. (It might not be obvious how to express the notion of a function, which appears in axiom (7). We can do this by regarding a function as a set of ordered pairs and then using Kuratowski's trick of encoding the ordered pair $\langle x, y \rangle$ as the set $\{\{x\}, \{x, y\}\}$.)

Zermelo-Fraenkel axioms

(1) *Extensionality.* Two sets are equal if they have the same elements.

(2) *Pairing.* For all x and y there exists a set whose elements are x and y.

(3) *Union.* For any set of sets there is a set which is their union.

(4) *Power set.* For all x there exists a set whose elements are the subsets of x.

(5) *Infinity.* There exists a set x such that $\emptyset \in x$ and for every u, if $u \in x$ then $u \cup \{u\} \in x$.

(6) *Foundation.* Every nonempty set has an $\in$-minimal element.

(7) *Choice.* Every set of nonempty sets has a choice function.

(8) *Separation scheme.* For all x there exists a set y whose elements are those $u \in x$ for which $\phi(u)$ holds.

(9) *Replacement scheme.* For all x, if for every $u \in x$ there is a unique v such that $\phi(u, v)$ holds, then there exists a set y obtained by replacing each $u \in x$ with the corresponding v.

In the separation and replacement schemes, ϕ can be any formula which contains no unquantified appearances of the variable y. If ϕ has other unquantified variables besides those indicated, then the resulting axiom should be understood as claiming that for any choice of these extra parameters the stated assertion holds.

Aside from extensionality and foundation, the Zermelo-Fraenkel axioms are all existence assertions which state that under certain conditions there is a set with certain properties. Moreover, choice is the only axiom for which the desired set is not explicitly identified. Thus, a crude apology for the axioms might go as follows. We start with full comprehension, the principle that for any formula ϕ there is a set consisting of all x such that $\phi(x)$ holds. We then realize that we can derive a contradiction in the form of Russell's paradox. We therefore replace full comprehension with various special cases which are sufficient to support a large amount of set-theoretic reasoning but do not give rise to any obvious contradictions.

Why do we need axiom schemes to express the principles of separation and replacement? Something similar happens with the induction scheme in PA. Induction is intuitively a single principle, and in Peano's original formulation it was expressed as a single axiom: any set of numbers which contains 0 and is stable under the successor operation contains every number. The drawback of this version is that it introduces a new type of object, a set of numbers, so that we now have to include a separate category of variables for this second type of object and write down axioms characterizing their behavior. This can be done; the result is known as *second order arithmetic.* Nowadays it is more common to replace the single second order induction axiom with a first order scheme, as we have done. This allows us to argue directly that all numbers have some property ϕ, as opposed to first forming the set of numbers which verify ϕ and then applying the second order induction axiom to this set.

In the case of set theory, in Zermelo's original formulation the separation principle also appeared as a single axiom stating that for any set x and any property P there exists a set whose elements are all the elements of x for which P holds. This too could be interpreted in second order terms, but again the more standard approach now is to replace the single reference to a "property" with a scheme of axioms, one for each formula which can be expressed in the language.

This can be a source of confusion if one develops an intuition for separation as the single assertion that for any set x and any formula ϕ we can form the set $\{u \in x \mid \phi(u)\}$. This single statement is *not* a theorem of ZFC and cannot even be directly expressed in the language of set theory. The same goes for the replacement scheme. We must think of ZFC as providing us the resources to make an infinite variety of separation arguments, and regard the fact that it does this as a theorem of the metatheory. Let us clarify this comment.

A proof in ZFC of some statement is nothing more than a finite string of symbols with certain elementary properties. It might not always be easy to tell whether a given statement is provable in ZFC — this is rather an understatement — but it is, in principle, trivial to check whether a given string of symbols constitutes a valid proof. One could easily write a computer program which would accomplish the latter task. Moreover, finite strings of symbols can be straightforwardly coded as natural numbers (using ASCII, for instance, to continue the programming theme), so we can recast statements that various strings are or are not proofs in ZFC as elementary assertions about natural numbers.

Thus, having encoded symbolic strings as natural numbers, we can reason within PA about the notion of provability in ZFC. For instance, we can prove in PA a single statement to the effect that all instances of the separation and replacement schemes are theorems of ZFC. In this picture ZFC is the *target theory* and PA is the *metatheory*.

We can now explain in general terms how our consistency and independence proofs are going to work. Suppose we are interested in establishing the assertion that the continuum hypothesis (CH) is consistent with ZFC, that is, its negation is not a theorem of ZFC. As we have just seen, this assertion can be put in arithmetical terms, so that we might hope to prove it in PA. But this goal is too ambitious. To see why, let us call a formal system *consistent* if no contradiction of the form $\psi \wedge \neg\psi$ is a theorem. Then we can say that a formula ϕ is consistent with ZFC in our original sense (its negation is not a theorem of ZFC) if and only if the formal system ZFC $+ \phi$ obtained by adding ϕ to ZFC as an additional axiom is consistent. For if $\neg\phi$ were a theorem of ZFC then $\phi \wedge \neg\phi$ would be a theorem of ZFC $+ \phi$; and conversely, if any contradiction were provable in ZFC $+ \phi$ then we could reinterpret this result as a proof in ZFC that assuming ϕ leads to a contradiction, which would amount to a proof of $\neg\phi$ in ZFC.

Thus, if we could prove in PA that some formula ϕ is consistent with ZFC then we would, as a byproduct, have proven in PA that ZFC itself is a consistent system. But we know for reasons related to Gödel's incompleteness theorem that this is impossible. (The problem is that it would imply that PA could prove its own consistency, since we can simulate PA within ZFC by interpreting the natural numbers set-theoretically.) In outline, this is how we know that there is no hope of proving straight consistency results for ZFC in PA.

In any case, the natural strategy for proving consistency results is set-theoretic, not number-theoretic. Just as one shows that the parallel postulate is consistent with the other Euclidean axioms by checking that they all hold in the Euclidean plane, we would like to show that CH is consistent with ZFC by constructing a *model* of ZFC + CH — a structure in which both the ZFC axioms and the continuum hypothesis hold. But models are set-theoretic objects. In PA we cannot even formulate the assertion that ZFC has a model, let alone prove this assertion.

Instead, we will adopt the following strategy. We start by defining a new formal system ZFC$^+$ which is, roughly, ZFC augmented by the assumption that there is a countable model $\mathbf{M}$ of ZFC. Working in ZFC$^+$, we then use the technique of forcing to convert $\mathbf{M}$ into a model $\mathbf{M}[G]$ of, say, ZFC $+$ CH. To the extent that we trust ZFC$^+$, this shows us that CH is consistent with ZFC. But regardless of what we think about ZFC$^+$, it will be possible to prove in PA that the consistency of ZFC implies the consistency of ZFC$^+$ and that the consistency of ZFC$^+$ implies the consistency of ZFC $+$ CH.

In short, a typical forcing argument is carried out in ZFC$^+$ and proves (in a schematic sense that we will discuss further) that there is a model $\mathbf{M}[G]$ of ZFC $+ \phi$, for some statement ϕ. Given that this result has been achieved in ZFC$^+$, we can then, working in PA, draw the conclusion that if ZFC is consistent then so is ZFC $+ \phi$. We will have proven a *relative consistency* result. This result is a theorem of the metatheory PA, although the bulk of the work was done within the auxiliary system ZFC$^+$.

Chapter 3

Well-Ordered Sets

In this and the next two chapters we will develop the basic theory of well-ordered sets, ordinals, and cardinals. Some of this material may already be familiar to the reader. We work in ZFC, but again, we are not concerned with the details of formalization. We simply ask the reader to accept that ordinary set-theoretic reasoning can be formalized in ZFC.

We should start by explaining how we interpret the natural numbers set-theoretically. The usual device is to encode them as sets by identifying each number with the set of all smaller numbers. Thus $0 = \emptyset$, $1 = \{0\} = \{\emptyset\}$, $2 = \{0, 1\} = \{\emptyset, \{\emptyset\}\}$, etc. In general $n + 1$ differs from n by the inclusion of one additional element, n itself, which is just to say that $n + 1 = n \cup \{n\}$.

The axiom of infinity entails that $\mathbb{N}$ exists. This is not quite immediate, because that axiom only asserts the existence of some set which contains $\emptyset$ and is stable under the successor operation $u \mapsto u \cup \{u\}$, whereas $\mathbb{N}$ is the minimal set with these properties. But the intersection of any family of sets with these properties again has these properties, and it follows that there exists a unique minimal set with these properties. This is how we define $\mathbb{N}$. The order relation is specified by setting $n < m$ just in case $n \in m$. This makes $\mathbb{N}$ well-ordered according to the following definition.

Definition 3.1. A *well-ordered set* is a totally ordered set with the property that any nonempty subset has a smallest element.

In general, if a well-ordered set W is nonempty then it has a smallest element x_0. If it contains more than one element then the subset $W \setminus \{x_0\}$ has a smallest element x_1, and so on. Thus the initial portion of W looks like a discrete sequence $x_0, x_1, \ldots$. If W is finite then this sequence terminates; otherwise, either the resulting infinite sequence (x_n) exhausts W or else the subset $W \setminus \{x_0, x_1, \ldots\}$ has a smallest element x_ω, and this might then be

9

followed by a next smallest element $x_{\omega+1}$, etc.

Well-ordering conveys a kind of temporal intuition of a set being built up one element after another, provided one is comfortable with the idea of transfinitely long constructions. Of course, we must be careful about invoking this kind of intuition when making rigorous arguments.

There is a simple alternative characterization of well-ordering that is also quite useful.

Proposition 3.2. *Let W be a totally ordered set. Then W is well-ordered if and only if it contains no strictly decreasing infinite sequence.*

Proof. ($\Rightarrow$) Suppose W contains a strictly decreasing infinite sequence. Then this sequence enumerates a subset of W with no smallest element, so W cannot be well-ordered.

($\Leftarrow$) Suppose W is not well-ordered and let A be a nonempty subset of W that contains no smallest element. Since A is nonempty we can choose $x_0 \in A$; then since x_0 is not the smallest element of A we can find $x_1 \in A$ satisfying $x_1 < x_0$; and repeating this process, we obtain a strictly decreasing infinite sequence in A. So W contains a strictly decreasing infinite sequence. $\square$

Although we stated Proposition 3.2 for totally ordered sets, it easily generalizes to partially ordered sets. The general conclusion is that a partially ordered set contains no strictly decreasing infinite sequence if and only if every nonempty subset has a minimal element, and the proof of this equivalence is virtually identical to the proof of Proposition 3.2. The same reasoning shows that the axiom of foundation is equivalent to the statement that there is no infinite sequence of sets (x_n) satisfying $x_0 \ni x_1 \ni x_2 \cdots$.

We now aim to prove a basic comparison result for well-ordered sets (Theorem 3.6).

Definition 3.3. Let W be a well-ordered set. An *initial segment* of W is a subset of the form $x^< = \{y \in W \mid y < x\}$ for some $x \in W$.

Note that we do not consider W to be an initial segment of itself.

For the rest of this chapter "isomorphism" means "order-isomorphism".

Lemma 3.4. *No well-ordered set is isomorphic to an initial segment of itself.*

Proof. Let W be a well-ordered set and let $x \in W$. Suppose $f : W \cong x^<$ is an isomorphism. Then $f(x) < x$, and applying f^n to both sides yields

$f^{n+1}(x) < f^n(x)$ for all n. Thus the sequence $x > f(x) > f^2(x) > \cdots$ is strictly decreasing, which contradicts Proposition 3.2. □

Lemma 3.5. *Suppose V and W are isomorphic well-ordered sets. Then the isomorphism between them is unique.*

Proof. Let $f, g : V \cong W$ be two isomorphisms between V and W. Then $h = g^{-1} \circ f$ is an isomorphism between V and itself. For any $x \in V$, the map h establishes an isomorphism between $x^<$ and $h(x)^<$, so Lemma 3.4 implies that $x = h(x)$. This shows that h must be the identity map, and hence that $f = g$. □

Although a function is by definition a set of ordered pairs, in normal mathematics this is not usually a particularly helpful way to think about functions. In set theory, however, it often is. The following proof will be the first of many instances in which we treat functions literally as sets.

Theorem 3.6. *Let V and W be well-ordered sets. Then exactly one of the following holds: V is isomorphic to an initial segment of W, W is isomorphic to an initial segment of V, or V and W are isomorphic. In all cases the isomorphism is unique.*

Proof. The second statement follows from Lemma 3.5. The fact that no more than one of the asserted cases can hold follows from Lemma 3.4.

Let

$$f = \left\{ \langle x, y \rangle \in V \times W \mid x^< \text{ is isomorphic to } y^< \right\}.$$

By Lemma 3.4, for each $x \in V$ there is at most one corresponding $y \in W$, and conversely, so that f is a bijection between a subset of V and a subset of W.

If $x^<$ is isomorphic to $y^<$ then it is clear that for any $x_0 < x$ there exists $y_0 < y$ such that $x_0^<$ and $y_0^<$ are isomorphic. This shows that f preserves order. It also shows that the domain of f is either V or an initial segment of V, and similarly its range is either W or an initial segment of W.

If either the domain of f is V or the range of f is W (or both) then we are done. So suppose both are initial segments, say $x^< \subset V$ and $y^< \subset W$. But then f is an isomorphism between $x^<$ and $y^<$, so that the pair $\langle x, y \rangle$ must belong to f, which is absurd. So the domain and range of f cannot both be initial segments. □

We conclude this chapter with Zermelo's proof that every set can be well-ordered. Given a set X, the idea of the proof is to use the axiom of choice to select one element from each nonempty subset of X and then use this information to build a well-ordering of X. Thus, our choice function first selects an element x_0 from X, then an element x_1 from $X \setminus \{x_0\}$, then an element x_2 from $X \setminus \{x_0, x_1\}$, and so on. We make this idea rigorous by considering all partial well-orderings of X which follow this pattern and showing that their union must be all of X. The key point is that a union of well-ordered sets must be well-ordered, if we assume that for any two of the sets one is an initial segment of the other.

Theorem 3.7. *(Well-ordering theorem) Every set can be well-ordered.*

Proof.   Let X be a set and let f be a choice function for the set of all nonempty subsets of X. Call a subset A of X a *z-set* if there is a well-ordering of A which has the property that $x = f(X \setminus x^<)$ for all $x \in A$. Thus $\emptyset$ is always a z-set, and if A is a z-set which is not all of X then $A \cup \{f(X \setminus A)\}$ is also a z-set (relative to the ordering that makes A with its given ordering an initial segment).

We claim that if A and B are isomorphic z-sets then $A = B$ and the isomorphism map is the identity. To see this, let $g : A \cong B$ be an isomorphism, assume g is not the identity map, and let x be the smallest element of A such that $x \neq g(x)$. Then $x^< = g(x)^<$, so that

$$x = f(X \setminus x^<) = f(X \setminus g(x)^<) = g(x),$$

a contradiction. We conclude that $g(x) = x$ for all $x \in A$, as claimed.

Since any initial segment of a z-set is itself a z-set, it now follows from Theorem 3.6 that if A and B are any z-sets, then either A is an initial segment of B, B is an initial segment of A, or $A = B$.

Let Z be the union of all the z-sets. By what we just said, there is a well-ordering on Z which makes it a z-set. But this means that Z must equal X, as otherwise $Z \cup \{f(X \setminus Z)\}$ would be a z-set, which is absurd. We conclude that X can be well-ordered.   □

The proof of Theorem 3.7 shows how a single global act of choice — selecting one element from each nonempty subset of x — can be used to implement a series of "dependent" choices where we first choose an element x_0 from X, then an element x_1 from $X \setminus \{x_0\}$, and so on. The reader may find it instructive at this point to return to the proof of the reverse direction of Proposition 3.2 and think about how a single global act of choice on the nonempty subsets of A could be used to define the decreasing sequence (x_n) constructed there with no further choices being necessary.

Chapter 4

Ordinals

An ordinal is a kind of canonical well-ordered set. Intuitively, any well-ordered set W can be converted into an ordinal by successively replacing each element $x \in W$ with the set $x^<$. Thus the first element of W is replaced with $\emptyset$, then the second element is replaced with $\{\emptyset\}$, then the third is replaced with $\{\emptyset, \{\emptyset\}\}$, and so on.

Recalling our definition of the natural numbers given in Chapter 3, this just amounts to saying that we are supposed to replace the $(n+1)$st element of W with the natural number n. However, as we will see in a moment, the ordinal construction continues to make sense at infinite stages.

The formal definition of ordinals goes as follows.

Definition 4.1. A set X is *transitive* if for any $x \in X$, every element of x is also an element of X. That is, $x \in X$ implies $x \subset X$. An *ordinal* is a transitive set that is totally ordered by $\in$.

Thus, given any two distinct elements of an ordinal, one must be an element of the other.

By the axiom of foundation, any set that is totally ordered by $\in$ must in fact be well-ordered by $\in$. So every ordinal is automatically well-ordered. The point of transitivity is that we are interested in the $\in$ relation among the elements of an ordinal, and we do not want these elements to contain any extraneous sets whose presence is irrelevant to this relation. The sense of this comment may become clearer in Chapter 6.

The $\in$ relation is irreflexive and antisymmetric by the axiom of foundation: we can never have $x \in x$, nor can we have both $x \in y$ and $y \in x$. Also, if α, β, and γ are ordinals satisfying $\alpha \in \beta$ and $\beta \in \gamma$, then $\alpha \in \gamma$ follows since γ is transitive. This shows that $\in$ is a transitive relation among ordinals. So it is an order relation, and we may accordingly write $\alpha < \beta$

13

instead of $\alpha \in \beta$, for any ordinals α and β.

Lemma 4.2. *Let α be an ordinal and let $\beta \in \alpha$. Then β is an ordinal and $\beta = \beta^<$.*

Proof. We begin with the second assertion. By definition, $\beta^<$ is the set of elements of α which precede, i.e., belong to, β. That is, $\beta^< = \alpha \cap \beta$. But $\beta \subset \alpha$ by transitivity, so we have $\beta^< = \beta$.

The fact that $\in$ totally orders β is an immediate consequence, since it totally orders α and $\beta = \beta^< \subset \alpha$. For transitivity, let $\gamma \in \beta$; then $\gamma \in \alpha$ since α is transitive, and therefore $\gamma = \gamma^< \subset \beta^< = \beta$. So β is transitive and totally ordered by $\in$, i.e., it is an ordinal. $\qquad\square$

In the next theorem, "isomorphic" means "order-isomorphic".

Theorem 4.3. *If α and β are ordinals then either $\alpha < \beta$, $\beta < \alpha$, or $\alpha = \beta$. Every well-ordered set is isomorphic to a unique ordinal.*

Proof. Let α and β be ordinals. Suppose first that α and β are isomorphic and let $f : \alpha \cong \beta$ be an isomorphism. If f is not the identity map then there is a smallest element $\gamma \in \alpha$ such that $f(\gamma) \neq \gamma$, but this leads to a contradiction because we then have

$$\gamma = \gamma^< = f(\gamma)^< = f(\gamma).$$

This shows that if α and β are isomorphic then they are equal. The first part of the theorem now follows from Theorem 3.6 together with the fact that any initial segment of an ordinal is an element of that ordinal.

For the second part, let W be a well-ordered set and let α be the set of all ordinals which are isomorphic to an initial segment of W. It follows from the first part of the theorem that this set (indeed, any set of ordinals) is totally ordered by $\in$, and if β is isomorphic to an initial segment of W it is clear that the same is true of any smaller ordinal, which implies that α is transitive. So α is an ordinal, and

$$f = \left\{ \langle x, \beta \rangle \in W \times \alpha \mid x^< \cong \beta^< \right\}$$

is an isomorphism between W and α by reasoning similar to that used in the proof of Theorem 3.6. (Its domain cannot be an initial segment of W because then α would meet the criterion for belonging to α.) The uniqueness statement follows from Lemma 3.4. $\qquad\square$

Together with Theorem 3.7 this result yields the following corollary.

Corollary 4.4. *Every set can be put in bijection with an ordinal.*

As we have already mentioned, in set theory we identify the natural numbers with the finite ordinals. Since the first infinite ordinal ω equals the set of all finite ordinals, it is identical to $\mathbb{N}$.

We define the sum $\alpha + \beta$ of two ordinals α and β to be the unique ordinal which is order-isomorphic to the well-ordered set formed by concatenating α and β. In particular, every ordinal α has an immediate successor $\alpha + 1$ whose elements are all the elements of α plus one additional element, namely α itself. That is, the equation $\alpha + 1 = \alpha \cup \{\alpha\}$ holds for all ordinals, not just the finite ones. An ordinal of the form $\alpha + 1$ is called a *successor ordinal*. A nonzero ordinal that is not a successor is called a *limit ordinal*, and the first example of such an ordinal is ω.

The principal feature of ordinals is that one can perform transfinite induction arguments on them. Although the general formulation of this fact is quite simple, it is worth emphasizing that the result is a "theorem scheme" rather than a single theorem. That is, for any formula ϕ in the language of set theory the following statement is provable in ZFC.

Theorem Scheme 4.5. *Suppose that for every ordinal α*

$$(\forall \beta < \alpha)\phi(\beta) \rightarrow \phi(\alpha)$$

holds. Then $(\forall \alpha)\phi(\alpha)$ holds.

Proof. Assume $\phi(\alpha)$ fails for some α. Then there is a smallest $\beta \leq \alpha$ for which $\phi(\beta)$ fails. However, this means that we have $(\forall \gamma < \beta)\phi(\gamma)$, so we can now use the hypothesis to infer $\phi(\beta)$, a contradiction. We conclude that $(\forall \alpha)\phi(\alpha)$. $\qquad \square$

We can prove in PA a single statement to the effect that every instance of Theorem Scheme 4.5 is a theorem of ZFC. This is done by formally writing down a proof template and verifying that for any choice of ϕ it specializes to a valid proof in ZFC. Notice that we must apply a different instance of the separation scheme in each case, at the step where we form the set of $\beta \leq \alpha$ for which ϕ fails.

The construction of the *cumulative hierarchy* (V_α) illustrates the idea of transfinite induction. We define the sets V_α recursively by letting V_α be the power set of $V_{<\alpha} = \bigcup_{\beta < \alpha} V_\beta$. (This is slightly different from the usual definition of V_α.) Thus $V_0 = \mathcal{P}(\emptyset) = \{\emptyset\}$, $V_1 = \mathcal{P}(\{\emptyset\}) = \{\emptyset, \{\emptyset\}\}$, etc. We can use Theorem Scheme 4.5 to prove that the V_α exist and are well-defined by taking $\phi(\alpha)$ to be the statement that there is a unique transfinite sequence $\{V_\beta \mid \beta \leq \alpha\}$ such that each V_β in the sequence is the power set of the union of the preceding V_γ. Since the formula $\phi(\alpha)$ makes sense for arbitrary α, this also shows how we can talk about all the V_α in a single assertion, despite the fact that the entire sequence (V_α) does not constitute a set — just as we can talk about all sets, or all ordinals.

The cumulative hierarchy exhausts the set-theoretic universe. Our proof of this fact will employ the *transitive closure* $TC(x)$ of a set x, which is defined to be the smallest transitive set containing x. It can be constructed as

$$TC(x) = x \cup x' \cup x'' \cup \cdots$$

where $y' = \bigcup_{u \in y} u$ denotes the union of all the elements of y. Also, note that the union $\bigcup_{u \in x} \alpha_u$ of any set of ordinals $\{\alpha_u \mid u \in x\}$ is an ordinal: it is transitive because it is a union of transitive sets, and it is totally ordered by $\in$ since it is a set of ordinals (Lemma 4.2) and the ordinals are totally ordered by $\in$ (Theorem 4.3). This union is in fact the least upper bound of the set $\{\alpha_u \mid u \in x\}$ and may or may not belong to it. If we want the least ordinal that is strictly greater than each α_u, we can take $\bigcup_{u \in x} (\alpha_u + 1)$.

Proposition 4.6. *If $\alpha \leq \beta$ then $V_\alpha \subseteq V_\beta$. Every set belongs to some V_α.*

Proof.   The first part follows from the observation that $X \subseteq Y$ implies that every subset of X is also a subset of Y, i.e., $\mathcal{P}(X) \subseteq \mathcal{P}(Y)$. So if $\alpha \leq \beta$ then $V_{<\alpha} \subseteq V_{<\beta}$ implies $V_\alpha \subseteq V_\beta$.

For the second part, let x be a set and start by assuming that every element u of x belongs to some V_α, say $u \in V_{\alpha_u}$. Then $\bar{\alpha} = \bigcup_{u \in x} (\alpha_u + 1)$ is an ordinal with the property that every element of x belongs to $V_{<\bar{\alpha}}$, which entails that x belongs to $V_{\bar{\alpha}}$. This shows that if every element of a set belongs to some V_α then the same is true of the set itself.

We conclude the proof by showing that every element of $TC(x)$, in particular every element of x, must belong to some V_α. Otherwise we could find an $\in$-minimal element u of $TC(x)$ among those which fail this condition. But since $TC(x)$ is transitive, this means that every element of u does belong to some V_α, so that u must belong to some V_α by what we just proved. This contradiction completes the proof. $\square$

It follows that for every α we have $V_\alpha = \bigcup_{\beta \leq \alpha} V_\beta$, and therefore at successor ordinals $V_{\alpha+1}$ is simply the power set of V_α. From this fact it is easy to show inductively that V_n has 2^n elements, for any $n \in \mathbb{N}$.

The smallest ordinal α such that $x \in V_\alpha$ is called the *rank* of x and denoted $\mathrm{rank}(x)$. It is not hard to see that the rank of any set is the smallest ordinal that exceeds the ranks of all of its elements, an observation which illuminates the nested hierarchical nature of the universe of sets.

Chapter 5

Cardinals

The finite ordinals all have distinct sizes: no two of them can be put in bijection with each other. This is no longer true of infinite ordinals; using the fact that a countable union of countable sets is countable, it is not hard to see that there are ordinals well beyond ω which can be put in bijection with ω. For instance, if we replace each element of ω with a distinct copy of ω, the result will be a countable well-ordered set whose corresponding ordinal, denoted ω^2, is greater than infinitely many limit ordinals. More generally, if each element of ω is replaced with any countable ordinal the result will still be countable and well-ordered. So we can concatenate ω copies of ω^2 to reach a countable ordinal ω^3, then concatenate ω copies of ω^3 to reach a countable ordinal ω^4, and so on. We can concatenate the sequence $\omega, \omega^2, \omega^3, \ldots$ to reach a countable ordinal denoted ω^ω. And so on.

On the other hand, since (as we will prove shortly) uncountable sets exist, Corollary 4.4 assures us that there must be uncountable ordinals. So evidently the chain of ordinals experiences intermittent jumps in size. The next definition singles out those ordinals at which these jumps take place.

Definition 5.1. A *cardinal* is an ordinal that cannot be put in bijection with any preceding ordinal.

According to Corollary 4.4, every set can be put in bijection with an ordinal. Clearly, the least ordinal that can be put in bijection with a given set X must be a cardinal. We call it the *cardinality of X* and denote it $\mathrm{card}(X)$.

The finite cardinals are the natural numbers $0, 1, 2, \ldots$. The smallest infinite cardinal $\omega = \mathbb{N}$ is also denoted $\aleph_0$, the next infinite cardinal after $\aleph_0$ is $\aleph_1$, the next one after that is $\aleph_2$, and so on. The supremum of the sequence $(\aleph_n)$ is denoted $\aleph_\omega$. This notation can be systematized by

17

observing that for any infinite cardinal κ, the set of all smaller infinite cardinals is well-ordered and therefore isomorphic to a unique ordinal α. The standard notation for κ is then $\aleph_\alpha$.

We can use the order relation on the ordinals to compare any two cardinals. Our first task is to show that this relation is detected by the existence of injective or surjective maps. We require an easy lemma.

Lemma 5.2. *Let α be an ordinal, let $W \subseteq \alpha$, and let β be the unique ordinal which is order-isomorphic to W with the order it inherits from α. Then $\beta \leq \alpha$.*

The proof of this lemma is a straightforward transfinite induction argument: let $f : W \cong \beta$ be an order-isomorphism and inductively infer that $f(\gamma) \leq \gamma$ for all $\gamma \in W$. This will verify that $\beta \subseteq \alpha$.

Proposition 5.3. *Let X and Y be sets. The following are equivalent:*

(i) $\mathrm{card}(X) \leq \mathrm{card}(Y)$
(ii) *there is an injective map from X into Y*
(iii) *there is a surjective map from Y onto X.*

Proof. (i) $\to$ (iii). Suppose $\mathrm{card}(X) \leq \mathrm{card}(Y)$. Since X is bijective with $\mathrm{card}(X)$ and Y is bijective with $\mathrm{card}(Y)$, it will suffice to show that there is a surjection from $\mathrm{card}(Y)$ onto $\mathrm{card}(X)$. This is trivial because $\mathrm{card}(Y)$ contains $\mathrm{card}(X)$.

(iii) $\to$ (ii). Suppose $f : Y \to X$ is a surjection. For each $x \in X$, let $g(x)$ be some element of $f^{-1}(x)$. Then g is an injection from X into Y.

(ii) $\to$ (i). Suppose there is an injection from X into Y. Then there is an injection from $\mathrm{card}(X)$ into $\mathrm{card}(Y)$. By the lemma, the range of this injection is in bijection with an ordinal $\beta \leq \mathrm{card}(Y)$. Since $\mathrm{card}(X)$ cannot be bijection with any ordinal which precedes it, this entails that $\mathrm{card}(X) \leq \mathrm{card}(Y)$. $\qquad\square$

The basic operations on cardinals are defined as follows.

Definition 5.4. Let λ and κ be cardinals. We define

(i) $\lambda + \kappa$ to be the cardinality of the disjoint union of λ and κ
(ii) $\lambda \cdot \kappa$ to be the cardinality of the cartesian product of λ and κ
(iii) λ^κ to be the cardinality of the set of all functions from κ to λ.

The cardinal sum of λ and κ need not agree with their ordinal sum defined in Chapter 4. For instance, the ordinal sum $\omega + 1$ is the successor of ω, whereas the cardinal sum $\aleph_0 + 1$ equals $\aleph_0$. But it should always be clear whether we are talking about ordinal or cardinal sums since we use different notation for ordinals (α, β, γ) and cardinals (λ, κ, θ). (Ordinal products and exponents also differ from cardinal products and exponents, but we have not defined these operations, and besides the brief mention of exponents at the start of this chapter we will not use them.)

We collect some elementary properties of cardinal arithmetic, with proofs left to the reader.

Proposition 5.5. *Let λ, κ, and θ be cardinals. Then*

(i) $\lambda + \kappa = \kappa + \lambda$
(ii) $\lambda + (\kappa + \theta) = (\lambda + \kappa) + \theta$
(iii) $\lambda \cdot \kappa = \kappa \cdot \lambda$
(iv) $\lambda \cdot (\kappa \cdot \theta) = (\lambda \cdot \kappa) \cdot \theta$
(v) $\lambda \cdot (\kappa + \theta) = \lambda \cdot \kappa + \lambda \cdot \theta$
(vi) $\lambda^{\kappa+\theta} = \lambda^\kappa \cdot \lambda^\theta$
(vii) $(\lambda \cdot \kappa)^\theta = \lambda^\theta \cdot \kappa^\theta$
(viii) $(\lambda^\kappa)^\theta = \lambda^{\kappa \cdot \theta}$.

We also have a simple formula for the cardinality of a power set:

Proposition 5.6. *Let X be a set. Then* $\mathrm{card}(\mathcal{P}(X)) = 2^{\mathrm{card}(X)}$.

This is true because the correspondence between a subset A of X and its characteristic function χ_A establishes a bijection between the power set of X and the set of all functions from X into the set $2 = \{0, 1\}$. Thus, it is the power set axiom that entails the existence of uncountable sets, via the following theorem of Cantor.

Theorem 5.7. *Let κ be a cardinal. Then* $\kappa < 2^\kappa$.

Proof. By Proposition 5.6, it will suffice to show that for any set X there is an injection from X into $\mathcal{P}(X)$, but no surjection from X onto $\mathcal{P}(X)$. The map $x \mapsto \{x\}$ settles the first point. For the second, let $f : X \to \mathcal{P}(X)$ be any function and define $A = \{x \in X \mid x \notin f(x)\}$. Thus $x \in A \leftrightarrow x \notin f(x)$, so we can never have $A = f(x)$. That is, A is not in the range of f. So there can be no surjection from X onto $\mathcal{P}(X)$. □

The other basic result about cardinal arithmetic is the following. Note that $\kappa^2 = \kappa \cdot \kappa = \mathrm{card}(\kappa \times \kappa)$ by Proposition 5.5 (vi).

Theorem 5.8. *Let κ be an infinite cardinal. Then* $\kappa = \kappa^2$.

Proof.   The inequality $\kappa \leq \kappa^2$ is easy. For the reverse inequality, suppose that we know $\lambda^2 < \kappa$ for every cardinal $\lambda < \kappa$. This will certainly be true for finite λ, and we may inductively assume that $\lambda^2 = \lambda < \kappa$ for all infinite $\lambda < \kappa$.

Now write $\kappa \times \kappa = \bigcup_{\alpha < \kappa} W_\alpha$ where α ranges over the ordinals less than κ and

$$W_\alpha = \big\{ \langle x, y \rangle \in \kappa \times \kappa \mid \max(x, y) = \alpha \big\}.$$

Fix a well-ordering of each W_α. (It is easy to write one down explicitly, but this is not necessary.) Then concatenate these well-orderings of the W_α to produce a well-ordering of $\kappa \times \kappa$. In this well-ordering, any initial segment is contained as a set in the square $\alpha \times \alpha$ for some ordinal $\alpha < \kappa$. Thus, by the induction hypothesis, every initial segment has cardinality strictly less than κ. So the corresponding ordinal (as in Theorem 4.3) has this property as well and therefore must be contained in κ. This shows that $\kappa^2 \leq \kappa$, as desired. $\square$

Corollary 5.9. *For any infinite cardinals λ and κ we have $\lambda + \kappa = \lambda \cdot \kappa = \max(\lambda, \kappa)$.*

Proof.   Without loss of generality suppose $\lambda \leq \kappa$. Then $\kappa \leq \lambda + \kappa \leq 2 \cdot \kappa \leq \lambda \cdot \kappa \leq \kappa^2 = \kappa$. $\square$

Corollary 5.10. *For any infinite cardinal κ we have $2^\kappa = \kappa^\kappa = (2^\kappa)^\kappa$.*

Proof. $2^\kappa \leq \kappa^\kappa \leq (2^\kappa)^\kappa = 2^{\kappa^2} = 2^\kappa$. $\square$

Corollary 5.9 trivializes the sum and product of infinite cardinals. The obvious next question is how to evaluate cardinal exponents, and the first problem here is to evaluate $2^{\aleph_0}$. We know from Theorem 5.7 that $2^{\aleph_0} \geq \aleph_1$. The *continuum hypothesis (CH)* is the assertion that $2^{\aleph_0} = \aleph_1$.

This completes our brief overview of elementary set theory. We now turn to the problem of proving consistency results.

Chapter 6

Relativization

In order to prove consistency results we propose to work with sets which play the role of miniature universes, in which the Zermelo-Fraenkel axioms hold alongside some additional statement such as the continuum hypothesis or its negation. We need to explain exactly what this means.

Let ϕ be a formula and let M be a set variable. The *relativization* of ϕ to M is the formula $\phi|_M$ obtained by restricting all quantifiers in ϕ to range over M. For instance, if ϕ is the pairing axiom then $\phi|_M$ says that for all x and y in M there is a set z in M such that for every u in M we have $u \in z \leftrightarrow (u = x \vee u = y)$. Intuitively, we evaluate the truth of $\phi|_M$ by ignoring all sets that do not appear in M.

The notion of relativization is accompanied by a range of terminology. Instead of asserting $\phi|_M$, we may informally say that ϕ *is true in* M or ϕ *holds in* M. We may equivalently assert that M *satisfies* ϕ or that M *models* ϕ, which is symbolically written as $M \models \phi$. However, it is important to understand that $\phi|_M$ is literally just another formula in the language of set theory, obtained from ϕ by inserting a clause about belonging to M after every quantifier, and that for our purposes these other expressions about satisfaction and modelling are all nothing more than synonyms of $\phi|_M$. When written out formally they are statements about M, not about ϕ. This needs to be emphasized because we are trying to clearly distinguish between theorems about sets, which are proven in ZFC, and theorems about formulas, which are proven in PA.

The way we affirm that the Zermelo-Fraenkel axioms hold in M is by asserting $\phi|_M$ for every axiom ϕ. We repeat the familiar warning that this cannot be accomplished with a single formula in the language of set theory, but will have to be done schematically.

We can easily identify the appropriate notion of isomorphism between sets that ensures they satisfy the same formulas. Say that a function f : $M \to N$ is an $\in$-*isomorphism* if it is a bijection and we have $x \in y \leftrightarrow f(x) \in f(y)$ for all $x, y \in M$.

Recall from Chapter 2 that the formal language of set theory only uses the logical symbols $\neg$, $\to$, and $\forall$. For any formula ϕ in the language of set theory with unquantified variables among $x_1, \ldots, x_n$, the following statement is provable in ZFC.

Proposition Scheme 6.1. *Let M and N be sets and suppose $f : M \cong N$ is an $\in$-isomorphism. Then for all $a_1, \ldots, a_n \in M$ we have*

$$\phi|_M(a_1, \ldots, a_n) \quad \leftrightarrow \quad \phi|_N(f(a_1), \ldots, f(a_n)).$$

Proof. If ϕ is the formula $x_1 \in x_2$ or $x_1 = x_2$ then the proposition is trivial. Now suppose we have proven the proposition for some formula ϕ. Then by appending to that proof the observation that $(\neg\phi)|_M$ is literally the same formula as $\neg(\phi|_M)$, and similarly for N, we can turn it into a proof of the proposition for the formula $\neg\phi$. If we have proofs of the proposition for ϕ and ψ then we can concatenate these proofs and add the observation that $(\phi \to \psi)|_M$ is literally the same formula as $\phi|_M \to \psi|_M$, and similarly for N, and thereby produce a proof of the proposition for $\phi \to \psi$. Finally, suppose we have a proof of the proposition for a formula ϕ with unquantified variables among $x_1, \ldots, x_n$. That is, we have a proof that

$$\phi|_M(a_1, \ldots, a_n) \quad \leftrightarrow \quad \phi|_N(f(a_1), \ldots, f(a_n))$$

for all $a_1, \ldots, a_n \in M$. From this plus the fact that f is a bijection we can deduce that for any $a_1, \ldots, \hat{a}_i, \ldots, a_n \in M$, using a hat to indicate that a_i is omitted, we have

$$(\forall x_i \in M)\phi|_M(a_1, \ldots, x_i, \ldots, a_n)$$

if and only if

$$(\forall x_i \in N)\phi|_N(f(a_1), \ldots, x_i, \ldots, f(a_n)).$$

So we get a proof of the proposition for the formula $(\forall x_i)\phi$. This shows how we build a proof of the proposition for any formula ϕ. □

In particular, if ϕ is a *sentence*, meaning that it has no unquantified variables, then the conclusion is simply $\phi|_M \leftrightarrow \phi|_N$.

The fact that each instance of Proposition Scheme 6.1 is a theorem of ZFC is itself a theorem of the metatheory. Its proof can be formalized in PA. It is interesting to note that the metatheoretic aspect of Theorem

Scheme 4.5 worked somewhat differently. In that case, in the metatheory we merely had to produce a proof template and check that after inserting any formula it becomes a proof in ZFC. In Proposition Scheme 6.1 the proofs grow in complexity as ϕ grows, and the metatheorem which states that every instance is a theorem of ZFC must be proven by induction on the complexity of ϕ.

Proposition Scheme 6.1 tells us that for the purpose of evaluating the truth of ϕ all that matters about M is the structure of the $\in$-relation among its elements. We found in Chapter 4 that transitivity is the right condition for expressing the idea that the elements of a set contain nothing superfluous with regard to the $\in$-relation. We will now show that there is a general technique for "collapsing" M to a transitive set. Say that M is *extensional* if any distinct $x, y \in M$ satisfy $x \cap M \neq y \cap M$. That is, if $x \neq y$ then there exists $u \in M$ which either belongs to x but not y or vice versa. This is just the relativization to M of the axiom of extensionality.

Lemma 6.2. *(Mostowski collapse lemma) Let M be an extensional set. Then there is a transitive set N and an $\in$-isomorphism $f : M \cong N$. The set N and the $\in$-isomorphism f are unique.*

Proof. We start by making the uniqueness argument. If N and f were not unique then there would be transitive sets N_1 and N_2 and an $\in$-isomorphism $g : N_1 \cong N_2$ which is not the identity map. Then $g(x) \neq x$ for some $x \in N_1$, and by the axiom of foundation we may suppose x is an $\in$-minimal element with this property. Now since N_1 and N_2 are transitive, every element of x belongs to N_1 and every element of $g(x)$ belongs to N_2; since g is an $\in$-isomorphism, it maps the elements of x to the elements of $g(x)$; and by minimality of x we have $g(u) = u$ for all $u \in x$. But this means that x and $g(x)$ have the same elements, i.e., $x = g(x)$, a contradiction. We conclude that g must be the identity map, which proves the uniqueness statement.

For existence, define a subset $A \subseteq M$ to be *downward closed* if $x \in A$ implies $u \in A$ for all $u \in x \cap M$; that is, every element of x that lies in M must belong to A. Then call a downward closed subset $A \subseteq M$ an *m-set* if it is $\in$-isomorphic to a transitive set. Now any downward closed subset of an m-set is an m-set, so the intersection of any two m-sets is an m-set; by the uniqueness statement we just proved, this shows that the maps from m-sets into transitive sets are compatible, so just as in the proof of Theorem 3.7 we can patch together all these maps and infer that the union Z of all the m-sets is an m-set. We need to use extensionality of M to ensure that when we patch maps together the resulting map is still injective. Now let $f : Z \cong N$ be the $\in$-isomorphism of Z with a transitive set. If $Z \neq M$ then we can find an $\in$-minimal element x of $M \setminus Z$ and extend f by setting $f(x) = \{f(u) \mid u \in x\}$. This contradicts maximality of Z and shows that we must have had $Z = M$. $\square$

This result may help to clarify the comment we made in Chapter 4 about wanting to exclude "extraneous" sets from ordinals. The collapse lemma gives us a systematic method for working our way up the elements of a set and discarding all extraneous sets, so that the end result is transitive.

The whole point of relativization is that the truth-value of ϕ in M might be different from its actual truth-value. For example, suppose that $\mathbb{N}$ and all of its subsets belong to M, but $\mathcal{P}(\mathbb{N})$ does not. Then the relativization of the power set axiom to M would be false. On the other hand, if M is transitive then there is a substantial class of formulas which, for any choice of parameters in M, have the same truth-value regardless of whether this is assessed inside or outside of M. We say that such formulas are *absolute* for M. For instance, it is clear that for any $x \in M$ the assertion "x is empty" holds in m if and only if x is actually empty, since the transitivity condition $x \subset M$ ensures that $(\exists u)(u \in x)$ is equivalent to $(\exists u \in M)(u \in x)$.

Generally speaking, any statement about x whose truth can be evaluated only by looking at the elements of x will be absolute for any transitive set. Restricting all quantifiers to range over M will not make any difference. We will return to this point in Chapter 8.

Chapter 7

Reflection

Recall that a sentence is a formula with no unquantified variables. In this chapter we will show that we can arrange for any true sentence to be true in some countable set M. It turns out that all we need to do is to include elements in M which verify certain existence statements. The following lemma scheme gives this construction.

Recall that the notation $a_1, \ldots, \hat{a}_j, \ldots, a_m$ indicates a list from which the jth entry has been omitted. For any finite list of formulas $\phi_1, \ldots, \phi_n$ whose unquantified variables are among $x_1, \ldots, x_m$, the following statement is provable in ZFC.

Lemma Scheme 7.1. *There is a nonempty countable set M such that for any $1 \leq i \leq n$, any $1 \leq j \leq m$, and any $a_1, \ldots, \hat{a}_j, \ldots, a_m \in M$,*

$$(\exists x_j)\phi_i(a_1, \ldots, x_j, \ldots, a_m) \quad \text{implies} \quad (\exists x_j \in M)\phi_i(a_1, \ldots, x_j, \ldots, a_m).$$

Proof. We start by defining an infinite sequence of sets (M_k). Let $M_0 = \{\emptyset\}$. Given M_k, we construct M_{k+1} as follows. First include all the elements of M_k in M_{k+1}. Next, for each $1 \leq i \leq n$, each $1 \leq j \leq m$, and each $a_1, \ldots, \hat{a}_j, \ldots, a_m \in M_k$, if there exists an x_j such that $\phi_i(a_1, \ldots, x_j, \ldots, a_m)$ holds then choose one such element and add it to M_{k+1}.

Since there are only countably many $(m-1)$-tuples of elements of a countable set, we inductively have that each M_k is countable. Let $M = \bigcup_{k=0}^{\infty} M_k$; this is a countable union of countable sets so it is also countable. Now suppose that for some $a_1, \ldots, \hat{a}_j, \ldots, a_m \in M$ there exists an x_j such that $\phi_i(a_1, \ldots, x_j, \ldots, a_m)$ holds. Then the elements $a_1, \ldots, \hat{a}_j, \ldots, a_m$ must all belong to some M_k and so we would have included such an element in M_{k+1}. This shows that whenever there exists some x_j such that $\phi_i(a_1, \ldots, x_j, \ldots, a_k)$, there exists such an x_j in M. $\square$

The statement of this lemma scheme is abbreviated. If an instance of it were actually written out in ZFC, we would have to assert a separate implication $(\exists x_j)\phi_i \to (\exists x_j \in M)\phi_i$ for each i and j.

Using the lemma scheme, we can prove the following theorem in ZFC, for any sentence ϕ.

Theorem Scheme 7.2. *(Reflection principle) There is a countable transitive set M such that $\phi \leftrightarrow \phi|_M$.*

Proof. Let ϕ' be the axiom of extensionality. Then we can produce a finite list of formulas $\phi_1, \ldots, \phi_n$ which includes ϕ and ϕ', and such that whenever $\neg\psi$ is in the list, so is ψ; whenever $\psi_1 \to \psi_2$ is in the list, so are ψ_1 and ψ_2; and whenever $(\forall x_j)\psi$ is in the list for some variable x_j, so is ψ. Apply Lemma Scheme 7.1 to the negations of the ϕ_i.

Let M be the countable set provided by the lemma scheme. For each ϕ_i we will prove

$$\phi_i(a_1, \ldots, a_m) \leftrightarrow \phi_i|_M(a_1, \ldots, a_m) \qquad (*)$$

for all $a_1, \ldots, a_m \in M$. This will be done first for the atomic ϕ_i, and then for progressively more complex ϕ_i.

If ϕ_i is atomic the proof of $(*)$ is trivial. Once $(*)$ is proven for ϕ_i it trivially follows for $\neg\phi_i$, and once it is proven for ϕ_{i_1} and ϕ_{i_2} it trivially follows for $\phi_{i_1} \to \phi_{i_2}$. Finally, suppose we have proven $(*)$ for a formula ϕ_i which contains the unquantified variable x_j. If for some choice of parameters in M we have $(\forall x_j)\phi_i(a_1, \ldots, x_j, \ldots, a_m)$ then we certainly have $(\forall x_j \in M)\phi_i(a_1, \ldots, x_j, \ldots, a_m)$, and hence, by hypothesis, $(\forall x_j \in M)\phi_i|_M(a_1, \ldots, x_j, \ldots, a_m)$. Conversely, if $(\forall x_j)\phi_i(a_1, \ldots, x_j, \ldots, a_m)$ fails then there must exist x_j such that $\neg\phi_i(a_1, \ldots, x_j, \ldots, a_m)$ holds, and by the lemma scheme some such x_j would have been included in M, so that $(\forall x_j \in M)\phi_i(a_1, \ldots, x_j, \ldots, a_m)$ would also fail; the conclusion that $(\forall x_j \in M)\phi_i|_M(a_1, \ldots, x_j, \ldots, a_m)$ fails then follows by hypothesis. Thus, once we have proven $(*)$ for ϕ_i we can prove it for $(\forall x_j)\phi_i$. So we can prove $(*)$ for each ϕ_i.

Since ϕ is among the ϕ_i and it contains no unquantified variables, we now know that $\phi \leftrightarrow \phi|_M$. And since ϕ' is among the ϕ_i, the set M is extensional, so we can invoke Lemma 6.2 and Proposition Scheme 6.1 to ensure transitivity. $\qquad \square$

Theorem Scheme 7.2 is known as a "reflection principle" because properties of the set-theoretic universe are reflected in M.

As a consequence of Theorem Scheme 7.2, we can prove, in PA, a metatheorem which states that if ϕ is a sentence which is provable in ZFC then ZFC also proves that there exists a countable transitive set which satisfies ϕ. This is a surprisingly strong result. For instance, ϕ could be the conjunction of any finite number of axioms of ZFC, and then any theorem provable from those axioms would also be provably true in M. So apparently countable sets suffice to support large amounts of set theory.

In particular, the sentence "there exist uncountable sets" is provable in ZFC and therefore also provably holds in some countable transitive set. This peculiar circumstance is called *Skolem's paradox*. This paradox is resolved by recognizing that there can be sets in M for which the assertion "x is countable" relativized to M is false, not because they are actually uncountable, but because no bijection with $\mathbb{N}$ appears in M.

We describe the formal system we will use for making forcing arguments. It is called ZFC$^+$. The language of ZFC$^+$ is the language of set theory with one additional constant symbol $\mathbf{M}$. The formulas are all the formulas of ZFC, plus any formula of ZFC with one of its unquantified variables replaced throughout by the symbol $\mathbf{M}$. The logical axiom schemes and the equality axioms are the same as for ZFC, as are the rules of inference. Note that in the logical scheme L5,

$$(\forall x)\phi(x) \rightarrow \phi(t),$$

the term t could be the constant symbol $\mathbf{M}$.

The non-logical axioms of ZFC$^+$ come in three groups:

- every non-logical axiom of ZFC
- the single assertion that $\mathbf{M}$ is countable and transitive
- the relativization of every non-logical axiom of ZFC to $\mathbf{M}$.

Thus, the axioms of ZFC$^+$ assert that $\mathbf{M}$ is a countable transitive set which satisfies the axioms of ZFC. However, the assertion that $\mathbf{M}$ models ZFC is not made as a single statement but as a scheme. The characterization we gave in Chapter 2 of ZFC$^+$ as "ZFC augmented by the assumption that there is a countable model $\mathbf{M}$ of ZFC" needs to be understood in this sense.

Recall that a formal system is consistent if no contradiction of the form $\phi \wedge \neg\phi$ is a theorem. The following metalemma is provable in PA.

Metalemma 7.3. *If ZFC is consistent then so is ZFC$^+$.*

Proof. Suppose ZFC$^+$ is inconsistent. Then there is a proof of $\phi \wedge \neg\phi$ for some formula ϕ. This proof involves only finitely many axioms, so in particular it involves only finitely many axioms of ZFC relativized to $\mathbf{M}$, say $\phi_1|_{\mathbf{M}}, \ldots, \phi_n|_{\mathbf{M}}$.

Since $\phi_1 \wedge \cdots \wedge \phi_n$ is a theorem of ZFC, according to Theorem Scheme 7.2 there is a proof in ZFC that some countable transitive set M satisfies $\phi_1 \wedge \cdots \wedge \phi_n$. Given such a proof, we can append to it a copy of the original proof of $\phi \wedge \neg\phi$ but with every occurrence of $\mathbf{M}$ replaced by M. The result will be a proof in ZFC of $\phi' \wedge \neg\phi'$ where ϕ' is ϕ with all occurrences of $\mathbf{M}$ replaced by M. Thus, we have shown that any inconsistency in ZFC$^+$ can be converted into an inconsistency in ZFC. □

In short, in order for ZFC$^+$ to be inconsistent, it would have to be the case that some finite set of ZFC axioms cannot consistently hold in a countable transitive set, but this would contradict the reflection principle.

In a typical forcing argument, we will work in ZFC$^+$ and enlarge the set $\mathbf{M}$ to another countable transitive set $\mathbf{M}[G]$ in a way that ensures that some formula, such as the continuum hypothesis, holds in $\mathbf{M}[G]$. We will also be able to prove that each axiom of ZFC holds in $\mathbf{M}[G]$. This will supply the premise of the following metatheorem, which is provable in PA.

Metatheorem 7.4. *Let ϕ be a sentence in the language of set theory. Suppose that in ZFC$^+$ we can define a set N and prove both the relativization of ϕ to N and the relativization of any axiom of ZFC to N. Then if ZFC is consistent, so is ZFC + ϕ.*

Proof. Saying that we can define N in ZFC$^+$ means that we can prove in ZFC$^+$ that there exists a set N which meets some defining condition. Now suppose there is a proof of a contradiction $\psi \wedge \neg\psi$ in ZFC + ϕ. Suppose this proof involves the Zermelo-Fraenkel axioms $\phi_1, \ldots, \phi_n$. Then in ZFC$^+$ we can first prove $(\phi \wedge \phi_1 \wedge \cdots \wedge \phi_n)|_N$, and then relativize the proof of $\psi \wedge \neg\psi$ to N. This yields a proof in ZFC$^+$ of $(\psi \wedge \neg\psi)|_N = \psi|_N \wedge \neg\psi|_N$, and by the metalemma this can then be converted into a proof of a contradiction in ZFC. So if ZFC is consistent, ZFC + ϕ must be consistent too. □

As we mentioned earlier, a theorem of the form "if ZFC is consistent then so is ZFC + ϕ" is called a relative consistency result. This is the kind of theorem we can prove in PA. But we have now done all the work we need to do in PA. The task which lies ahead is to construct $\mathbf{M}[G]$ and show that it satisfies both the Zermelo-Fraenkel axioms and some other additional statement or statements. All of this work will be carried out in ZFC$^+$.

Chapter 8

Forcing Notions

From this point on, in chapters which develop the theory of forcing we will work in ZFC$^+$, and in chapters focused on applications we will work in ZFC. The next several chapters take place in ZFC$^+$.

In the world of ZFC$^+$, we have a countable transitive set $\mathbf{M}$ in which the Zermelo-Fraenkel axioms are true. This means that we can relativize to $\mathbf{M}$ any construction that makes sense in ZFC. For instance, we can carry out, in $\mathbf{M}$, the usual constructions of the set of natural numbers, the power set of the natural numbers, and the first uncountable cardinal. Let us call the results of these constructions $\mathbb{N}^{\mathbf{M}}$, $\mathcal{P}(\mathbb{N})^{\mathbf{M}}$, and $\aleph_1^{\mathbf{M}}$. What are these sets really, from the point of view of ZFC$^+$?

The concept of absoluteness mentioned at the end of Chapter 6 is helpful here. Consider the statement "the elements of x are totally ordered by $\in$". The only sets relevant to its truth are the elements of x, and if $x \in \mathbf{M}$ then all these elements belong to $\mathbf{M}$ by transitivity, so relativizing to $\mathbf{M}$ has no effect. $\mathbf{M}$ satisfies "the elements of x are totally ordered by $\in$" if and only if the elements of x are totally ordered by $\in$, for any $x \in \mathbf{M}$. Likewise, the statement "x is transitive" only quantifies over elements of x and elements of elements of x, so it too is absolute for $\mathbf{M}$. Combining these two conditions shows that being an ordinal is absolute. That is, if $x \in \mathbf{M}$ then $\mathbf{M}$ satisfies "x is an ordinal" if and only if x is an ordinal. By similar reasoning, the property of being the smallest limit ordinal is also absolute for $\mathbf{M}$, and this shows that $\mathbb{N}^{\mathbf{M}} = \mathbb{N}$. The set that plays the role of the natural numbers in $\mathbf{M}$ is the actual set of natural numbers.

On the other hand, since $\mathbf{M}$ is transitive and countable, every set that appears in $\mathbf{M}$ must also be countable, so we know that $\mathcal{P}(\mathbb{N})$ and $\aleph_1$ cannot belong to $\mathbf{M}$. Instead, $\mathcal{P}(\mathbb{N})^{\mathbf{M}}$ is the set of all subsets of $\mathbb{N}$ that appear in $\mathbf{M}$, i.e., $\mathcal{P}(\mathbb{N})^{\mathbf{M}} = \mathcal{P}(\mathbb{N}) \cap \mathbf{M}$, and $\aleph_1^{\mathbf{M}}$ is the smallest infinite ordinal with the property that no bijection between it and ω appears in $\mathbf{M}$. However, $\aleph_1^{\mathbf{M}}$ is in fact a countable ordinal.

Now since $\mathcal{P}(\mathbb{N})^{\mathbf{M}}$ and $\aleph_1^{\mathbf{M}}$ are both countable, there must be a bijection between them. We just do not know whether such a bijection appears in $\mathbf{M}$. But if there is no bijection between $\mathcal{P}(\mathbb{N})^{\mathbf{M}}$ and $\aleph_1^{\mathbf{M}}$ in $\mathbf{M}$, we could try to convert $\mathbf{M}$ into a model of ZFC + CH by enlarging it so as to include one. There are two hurdles to overcome in order for this idea to work: first, we have to figure out what other sets need to be included along with the new bijection in order to ensure that the Zermelo-Fraenkel axioms still hold; and second, we have to show that the continuum hypothesis holds in this new model. The second point is slightly subtle and we will return to it later. For now we concentrate on the problem of enlarging $\mathbf{M}$ in such a way that the Zermelo-Fraenkel axioms continue to hold. There is a general body of machinery for accomplishing this.

In this general setup we aim to add a certain kind of object to $\mathbf{M}$. Instead of immediately specifying this object, we start by identifying a family of possible partial constructions of it which can be performed in $\mathbf{M}$. Ordered by inclusion, this family becomes a poset in which later stages of a potential construction lie above earlier stages. Thus, we carry out the construction as we move up the poset. The idea is that not all constructions which can be realized in this way lead to models of ZFC, but those which are "generic" relative to $\mathbf{M}$ do.

Definition 8.1. Let P be a set.

(a) If $p, q \in P$ and $q \supseteq p$ then q is an *extension* of p.
(b) Two elements $p, q \in P$ are *compatible* if they have a common extension in P.
(c) A subset D of P is *dense* if every $p \in P$ has an extension in D. It is *dense above p* if every extension of p has an extension in D.

A *forcing notion* for $\mathbf{M}$ is any nonempty set $P \in \mathbf{M}$.

Definition 8.2. An *ideal* of a forcing notion P is a subset $G \subseteq P$ which satisfies

(i) if $q \in G$ and $q \supseteq p \in P$ then $p \in G$ (*downward stability*)
(ii) if $p_1, p_2 \in G$ then there exists $q \in G$ with $p_1, p_2 \subseteq q$ (*directedness*).

It is *generic* relative to $\mathbf{M}$ if it intersects every dense subset $D \subseteq P$ that lies in $\mathbf{M}$.

An example may be helpful at this point. Suppose the desired object we wish to add to **M** is a function from $\mathbb{N}$ into $\{0,1\}$. A partial construction of this object might be a function from a finite subset A of $\mathbb{N}$ into $\{0,1\}$. Let P be the set of all such *finite partial functions*. If $f, g \in P$ then the relation $f \subseteq g$, regarding f and g as sets of ordered pairs, just means that g extends f in the usual sense of extension of a function. Also, two functions in P have a common extension in P precisely if they agree on their common domain, i.e., they are compatible as functions. This example should help explain the terminology introduced in Definition 8.1 (a) and (b).

A crucial point that deserves emphasis is that forcing notions must belong to **M**. This condition could be met by describing some construction of a set and then relativizing that construction to **M**. In the preceding example this is not necessary, because the construction we gave there produces the same result inside and outside **M**. This is not too hard to see, given that we already know $\mathbb{N}^{\mathbf{M}} = \mathbb{N}$. First we claim that the property of being a function from a finite subset of $\mathbb{N}$ into $\{0,1\}$ is absolute for **M**. We can determine whether f is a set of ordered pairs without looking outside of $TC(f)$, so this condition is absolute, and we can determine whether a subset of $\mathbb{N}$ is finite by checking whether it has a greatest element, so this condition is also absolute. The claim now follows straightforwardly. Thus, within **M** we can correctly identify which of its elements are finite partial functions from $\mathbb{N}$ into $\{0,1\}$. In order to show that "finite partial function from $\mathbb{N}$ into $\{0,1\}$" means the same thing inside and outside **M**, we also need to check that every finite partial function from $\mathbb{N}$ into $\{0,1\}$ belongs to **M**; this is true because any such function can be constructed using a finite number of applications of the pairing axiom, and relativizing this construction to **M** changes nothing.

Now let G be an ideal of the set of finite partial functions from $\mathbb{N}$ to $\{0,1\}$. Since any two elements of an ideal must be compatible, this means that we can patch together (take the union of) all the functions that appear in G to produce a single function from a subset of $\mathbb{N}$ into $\{0,1\}$. If G is generic then we may further argue that the domain of this function must be all of $\mathbb{N}$. To see this, let $k \in \mathbb{N}$ be arbitrary and let D_k be the set of functions in P whose domain includes k. This is a dense subset because any function in P can be extended so that its domain contains k. So any generic ideal will have to meet every D_k, i.e., for each k it must include a function whose domain contains k. Thus, when we patch all of these functions together, we get a total function from $\mathbb{N}$ into $\{0,1\}$. This illustrates the idea that a generic ideal represents a complete construction of the object whose partial constructions are represented by elements of the forcing notion. The sets D_k in this example lie in $\mathbf{M}$, since we do not have to look outside the transitive closure of a function to determine whether k lies in its domain. This is important because we do not require a generic ideal to meet every dense set, only the dense sets appearing in $\mathbf{M}$.

We are always able to construct generic ideals by going outside of $\mathbf{M}$.

Theorem 8.3. *Let P be a forcing notion and let $p_0 \in P$. Then there is a generic ideal of P that contains p_0.*

Proof. Since $\mathbf{M}$ is countable, we can enumerate all the dense subsets of P that lie in $\mathbf{M}$. Let (D_n) be such an enumeration. Construct an infinite sequence (p_n) by, for each n, choosing p_{n+1} to be an extension of p_n that lies in D_n. This is possible since each D_n is dense. Finally, let

$$G = \big\{ p \in P \mid p \subseteq p_n \text{ for some } n \big\}.$$

It is immediate that G satisfies condition (i) of Definition 8.2 and that it meets every D_n. For property (b), let $p, q \in G$; since the sequence (p_n) is increasing, we can find n such that p_n contains both p and q. But then p_n is itself a common extension of p and q. Thus G is a generic ideal. $\square$

If G is a generic ideal of the set of all finite partial functions from $\mathbb{N}$ into $\{0, 1\}$, then as we saw above, the union of all the functions in G will be a total function $\tilde{f}$ from $\mathbb{N}$ into $\{0, 1\}$. Although $\mathbf{M}$ already contains plenty of functions from $\mathbb{N}$ to $\{0, 1\}$, we claim that $\tilde{f}$ cannot belong to $\mathbf{M}$. For if $\tilde{g}$ is any function from $\mathbb{N}$ to $\{0, 1\}$ that lies in $\mathbf{M}$, then, working in $\mathbf{M}$, we can form the set $D_{\tilde{g}}$ of all finite partial functions which disagree with $\tilde{g}$ at some point in their domain. It is easy to see that this set is dense: any finite partial function can be extended to one new element $k \in \mathbb{N}$ and assigned a value at k that differs from $\tilde{g}(k)$. So $D_{\tilde{g}}$ is a dense set that lies in $\mathbf{M}$, and hence G must meet $D_{\tilde{g}}$, which implies that $\tilde{f}$ cannot equal $\tilde{g}$. Thus, in this case we *must* go outside of $\mathbf{M}$ in order to construct a generic ideal. Requiring G to meet every dense set in $\mathbf{M}$ prevents us from being able to "predict" the construction of G from within $\mathbf{M}$.

Chapter 9

Generic Extensions

We continue to work in ZFC$^+$. Throughout this chapter, we fix a forcing notion P and a generic ideal G of P.

The point we ended the last chapter on can be made more generally. Say that P is *trivial* if it contains an element any two of whose extensions are compatible. If such an element exists, then it is easy to see that the elements which are compatible with it constitute a generic ideal. On the other hand, if P is nontrivial then generic ideals cannot be constructed in **M**. This is because nontriviality implies that the complement of any ideal must be dense, since any element has a pair of incompatible extensions, only one of which can belong to the ideal. So if G belonged to **M** then its complement $P \backslash G$ would be a dense set in **M**, contradicting the requirement that G must meet every dense set in **M**.

Thus, G typically does not belong to **M**. Our main goal in this chapter is to work out a way of augmenting **M** by G. We want to ensure that the augmented set **M**[G] still models ZFC, and so we have to determine which sets should be included in **M**[G] in order to make this happen.

Let us begin by analyzing how the cumulative hierarchy construction relativizes to **M**. It is easiest to do this in terms of the associated concept of rank. Since the property of being an ordinal is absolute for **M**, the rank of any $x \in$ **M**, as defined in **M**, will be an ordinal; call it rank$^{\mathbf{M}}(x)$. But both inside and outside of **M**, the rank of x is the smallest ordinal that exceeds the ranks of all of its elements, so an easy induction shows that rank$^{\mathbf{M}}(x) = $ rank(x) for all $x \in$ **M**. This shows that x belongs to $V_\alpha^{\mathbf{M}}$ if and only if it belongs to V_α, and thus $V_\alpha^{\mathbf{M}} = V_\alpha \cap$ **M** for all $\alpha \in$ **M**.

M is both shorter and thinner than the entire universe. It is "shorter" because not all ordinals belong to **M**, only those less than a certain count-able ordinal (namely, the supremum of all the ordinals in **M**). It is "thinner" because for $\alpha \geq \omega$, $V_\alpha^{\mathbf{M}}$ does not contain every subset of $V_{<\alpha}^{\mathbf{M}}$, only those (countably many) subsets which appear in $\mathbf{M}$.

In order to define $\mathbf{M}[G]$, we will modify the construction of the $V_\alpha^{\mathbf{M}}$ in a way that incorporates the extra information contained in G. The idea is that being allowed to consult G should sometimes enable us to find more subsets of a given set X than we could when working in $\mathbf{M}$, and this can be used to enrich the construction of the cumulative hierarchy.

One obvious way to use G to obtain new subsets of X is by considering functions $f : X \to P$ and forming the sets $f^{-1}(G) \subseteq X$. More generally, given any relation $R \subseteq X \times P$ we can form the *relational inverse*

$$R^{-1}(G) = \big\{ x \in X \mid \langle x, p \rangle \in r \text{ for some } p \in G \big\}.$$

This suggests how we might modify the construction of a single $V_\alpha^{\mathbf{M}}$ by using relations which appear in $\mathbf{M}$ to extract information from G. But it does not straightforwardly generalize to successive steps, when we will need to form subsets of newly introduced sets that do not belong to $\mathbf{M}$. The solution to this difficulty is to separate the process into two parts — exhibiting the relation R, and using it to form the set $R^{-1}(G)$ — and to perform the first step globally before passing to the second. Thus we recursively build up an entire hierarchy of relations, and after this construction is complete we then apply a second recursion to convert these relations into sets in $\mathbf{M}[G]$. A benefit of this approach is that the first recursion takes place entirely in $\mathbf{M}$. Thus, by regarding the relation as a "name" for the associated set it is used to construct, we will be able talk about the elements of $\mathbf{M}[G]$ within $\mathbf{M}$. The meaning of this comment should become clearer as we proceed.

Definition 9.1. Recursively define the *P-name hierarchy* (N_α), with α ranging over the ordinals in $\mathbf{M}$, by letting N_α consist of all the sets $\tau \in \mathbf{M}$ with the properties that

(i) every element of τ is an ordered pair $\langle \sigma, p \rangle$ with $\sigma \in N_{<\alpha}$ and $p \in P$,
   i.e., $\tau \subseteq N_{<\alpha} \times P = \left(\bigcup_{\beta < \alpha} N_\beta \right) \times P$
(ii) if $\langle \sigma, p \rangle \in \tau$ then $\langle \sigma, q \rangle \in \tau$ for every extension q of p.

The sets τ are called *P-names*, and the least α such that $\tau \in N_\alpha$ is the *name rank* of τ.

We will say something about the significance of condition (ii) in a moment. Note that all the quantification in Definition 9.1 is restricted to $\mathbf{M}$; in other words, this definition has already been relativized to $\mathbf{M}$.

P-names can be "evaluated" using the generic ideal G.

Definition 9.2. The *value* τ^G of a P-name τ is defined recursively on name rank by

$$\tau^G = \left\{ \sigma^G \mid \sigma \in \tau^{-1}(G) \right\}$$

where $\tau^{-1}(G)$ is the relational inverse discussed above. The *generic extension* $\mathbf{M}[G]$ of $\mathbf{M}$ associated to G is the set of all values of P-names,

$$\mathbf{M}[G] = \left\{ \tau^G \mid \tau \text{ is a } P\text{-name} \right\}.$$

Each pair $\langle \sigma, p \rangle$ in τ can be thought of as representing a potential element σ^G of τ^G, one which is realized as an actual element only if p belongs to G. We can now see that condition (ii) in Definition 9.1 is merely cosmetic. If $\langle \sigma, p \rangle \in \tau$ then for any extension q of p the pair $\langle \sigma, q \rangle$ does not independently contribute to τ^G, because $q \in G$ implies $p \in G$.

If $\tau_0 \in \mathbf{M}$ is any set of pairs $\langle \sigma, p \rangle$ with σ a P-name and $p \in P$, we define the P-name *generated by* τ_0 to be the set τ of pairs $\langle \sigma, q \rangle$ such that $p \subseteq q \in P$ for some $\langle \sigma, p \rangle \in \tau_0$. By the previous comment we have $\tau^G = \tau_0^G$ when Definition 9.2 is extended to τ_0 in the obvious way.

Next, we verify that $\mathbf{M} \subseteq \mathbf{M}[G]$ and $G \in \mathbf{M}[G]$. We do this using *canonical names* $\check{x}$ and Γ for $x \in \mathbf{M}$ and G.

Definition 9.3. Define a P-name $\check{x}$ for each $x \in \mathbf{M}$ recursively on the rank of x by the formula

$$\check{x} = \left\{ \langle \check{u}, p \rangle \mid u \in x \text{ and } p \in P \right\}$$

and then define the P-name Γ by the formula

$$\Gamma = \left\{ \langle \check{p}, q \rangle \mid p, q \in P \text{ and } p \subseteq q \right\}.$$

That is, Γ is the P-name generated by $\left\{ \langle \check{p}, p \rangle \mid p \in P \right\}$.

The idea is that $\check{x}$ tags all the elements of x with every element of P in order to ensure that they all appear in $\check{x}^G$, while Γ tags each element of P with itself, so that Γ^G only picks up the elements of G.

Proposition 9.4. *We have* $\check{x}^G = x$ *for all* $x \in \mathbf{M}$ *and* $\Gamma^G = G$. *Thus* $\mathbf{M} \subseteq \mathbf{M}[G]$ *and* $G \in \mathbf{M}[G]$. $\mathbf{M}[G]$ *is countable and transitive. It contains the same ordinals as* $\mathbf{M}$.

Proof. The first assertion is proven inductively on $\text{rank}(x)$. First observe that $\check{x}^{-1}(G) = \left\{ \check{u} \mid u \in x \right\}$. So assuming $\check{u}^G = u$ holds for sets of all smaller ranks, we have

$$\check{x}^G = \left\{ \check{u}^G \mid u \in x \right\} = \left\{ u \mid u \in x \right\} = x.$$

We conclude that $\check{x}^G = x$ for all $x \in \mathbf{M}$. Next, observe that $\Gamma^{-1}(G) = \{\check{p} \mid p \in G\}$. So

$$\Gamma^G = \{\check{p}^G \mid p \in G\} = \{p \mid p \in G\} = G.$$

It immediately follows that $\mathbf{M} \subseteq \mathbf{M}[G]$ and $G \in \mathbf{M}[G]$. $\mathbf{M}[G]$ is countable because the set of all P-names is a subset of $\mathbf{M}$ and hence is countable. It is transitive because, by definition, the elements of the value of any P-name are themselves all values of P-names.

Finally, an easy induction shows that the rank of τ^G is no greater than the name rank of τ. So the rank of any set in $\mathbf{M}[G]$ is less than or equal to some ordinal in $\mathbf{M}$. But the rank of any ordinal equals itself, so this shows that no new ordinals are intoduced in $\mathbf{M}[G]$. $\qquad\square$

We want to verify that all the Zermelo-Fraenkel axioms are true in $\mathbf{M}[G]$. This is going to take a bit of preparation, but some parts are easy and can be proven now.

Theorem 9.5. *The axioms of extensionality, pairing, union, infinity, and foundation are true in* $\mathbf{M}[G]$.

Proof. Extensionality follows from the fact that $\mathbf{M}[G]$ is transitive. Pairing holds because if τ_1 and τ_2 are any two P-names then the P-name $\mathrm{up}(\tau_1, \tau_2)$ ("unordered pair") consisting of all pairs $\langle \sigma, p \rangle$ with $\sigma = \tau_1$ or τ_2 and $p \in P$ is a P-name whose value is $\{\tau_1^G, \tau_2^G\}$.

For the union axiom, let τ be any P-name. Then

$$\tilde{\tau} = \left\{ \langle \pi, p \rangle \mid \text{for some } \sigma \text{ we have } \langle \pi, p \rangle \in \sigma \text{ and } \langle \sigma, p \rangle \in \tau \right\}$$

is a P-name; we will verify that it evaluates to the union of the sets in τ^G. That is, we must show that every element of $\tilde{\tau}^G$ is an element of an element of τ^G, and vice versa. In one direction, let $x \in \tilde{\tau}^G$. Then $x = \pi^G$ for some $\langle \pi, p \rangle \in \tilde{\tau}$ with $p \in G$. Since $\langle \pi, p \rangle \in \tilde{\tau}$, we can find σ such that $\langle \pi, p \rangle \in \sigma$ and $\langle \sigma, p \rangle \in \tau$; then we have both $x = \pi^G \in \sigma^G$ and $\sigma^G \in \tau^G$. So x belongs to an element of τ^G. Conversely, let $x \in \sigma^G$ for some $\langle \sigma, p \rangle \in \tau$ with $p \in G$. Then $x = \pi^G$ for some $\langle \pi, q \rangle \in \sigma$ with $q \in G$. Choosing $r \supseteq p, q$ in G, we have $\langle \pi, r \rangle \in \sigma$ and $\langle \sigma, r \rangle \in \tau$, so that $\langle \pi, r \rangle \in \tilde{\tau}$ and hence $x = \pi^G \in \tilde{\tau}^G$. We conclude that $\tilde{\tau}$ evaluates to the union of the sets in τ^G.

Infinity follows from the fact that $\mathbf{M} \subseteq \mathbf{M}[G]$. Foundation is trivial; it holds in any set. $\qquad\square$

For future use, we note that $\mathrm{op}(\tau_1, \tau_2) = \mathrm{up}(\mathrm{up}(\tau_1, \tau_1), \mathrm{up}(\tau_1, \tau_2))$ ("ordered pair") is a P-name whose value is $\langle \tau_1^G, \tau_2^G \rangle$.

Chapter 10

Forcing Equality

We fix a forcing notion P but do not fix a generic ideal. Now for any P-names $\tau_1, \ldots, \tau_n$ and any formula ϕ in the language of set theory with unquantified variables among $x_1, \ldots, x_n$, it makes sense to ask, in ZFC$^+$, for which generic ideals G the assertion $\phi(\tau_1^G, \ldots, \tau_n^G)$ holds in $\mathbf{M}[G]$. Is it ever possible to infer the truth-value of $\phi|_{\mathbf{M}[G]}(\tau_1^G, \ldots, \tau_n^G)$ directly from some structural property of G? Conceivably, there could be cases where the mere presence of a single element $p \in P$ in G could be enough to ensure that $\phi(\tau_1^G, \ldots, \tau_n^G)$ holds in $\mathbf{M}[G]$. In other words, it could happen that $\mathbf{M}[G] \models \phi(\tau_1^G, \ldots, \tau_n^G)$ for any generic ideal G that contains p. If so, we say that p *forces* $\phi(\tau_1, \ldots, \tau_n)$ and write

$$p \Vdash \phi(\tau_1, \ldots, \tau_n).$$

For each formula ϕ, the assertion that p forces $\phi(\tau_1, \ldots, \tau_n)$ is a statement about p and the τ_i which can be made in the language of ZFC$^+$.

The forcing relation might appear impossibly strong — merely knowing that G contains p would not seem to tell us much — but it is not. The fundamental theorem of forcing (really a theorem scheme) effectively states that *everything* which is true in $\mathbf{M}[G]$ is forced by some $p \in G$, and furthermore, information about which p force which $\phi(\tau_1, \ldots, \tau_n)$ is available in $\mathbf{M}$. This is an extremely powerful result and it is the cornerstone of the theory of forcing.

The hardest case of the fundamental theorem is actually the simplest formula, the atomic formula $x_1 = x_2$. We treat this case first. Denote the max of the name ranks of σ_1 and σ_2 by $nr(\sigma_1, \sigma_2)$. Now recursively define a transfinite sequence of sets $(\mathcal{F}_\alpha)$, for α an ordinal in $\mathbf{M}$, by letting $\mathcal{F}_\alpha$ be the set of ordered triples $\langle p, \tau_1, \tau_2 \rangle \in P \times N_\alpha^2$ which satisfy the two conditions

(i) for every $\langle \sigma_1, q_1 \rangle \in \tau_1$ with $q_1 \supseteq p$, there exists $\langle \sigma_2, q_2 \rangle \in \tau_2$ such that $q_2 \supseteq q_1$ and $\langle q_2, \sigma_1, \sigma_2 \rangle \in \mathcal{F}_{nr(\sigma_1, \sigma_2)}$

(ii) for every $\langle \sigma_2, q_2 \rangle \in \tau_2$ with $q_2 \supseteq p$, there exists $\langle \sigma_1, q_1 \rangle \in \tau_1$ such that $q_1 \supseteq q_2$ and $\langle q_1, \sigma_1, \sigma_2 \rangle \in \mathcal{F}_{nr(\sigma_1, \sigma_2)}$.

Note that the recursion makes sense because $nr(\sigma_1, \sigma_2) < \alpha$ — the name rank of any P-name is strictly greater than the name rank of any of its elements. We will show that $\langle p, \tau_1, \tau_2 \rangle$ belongs to $\mathcal{F}_\alpha$ if and only if p forces $\tau_1 = \tau_2$. The proof is inductive because we have to use this fact for σ_1 and σ_2 when proving it for τ_1 and τ_2.

The idea behind condition (i) is that we are trying to ensure every element of τ_1^G also lies in τ_2^G, for any generic ideal G that contains p. Now every element of τ_1^G is of the form σ_1^G for some $\langle \sigma_1, q_1 \rangle \in \tau_1$ with $q_1 \in G$, and by replacing q_1 with a common extension of p and q_1 we can assume $q_1 \supseteq p$. We want to ensure σ_1^G belongs to τ_2^G, and we do this with a pair $\langle \sigma_2, q_2 \rangle \in \tau_2$ such that q_2 forces $\sigma_1 = \sigma_2$. Since any $q_1' \supseteq q_1$ also satisfies $\langle \sigma_1, q_1' \rangle \in \tau_1$, we will see that condition (i) guarantees there are sufficiently many such q_2 that one of them must belong to G. Thus, σ_2^G appears in τ_2^G, and it is forced by q_2 to equal σ_1^G. In the same way, condition (ii) ensures that every element of τ_2^G also lies in τ_1^G.

The technique of replacing q_1 with a common extension of p and q_1 so that we can assume $q_1 \supseteq p$ is so useful that it deserves special terminology. The point is that $\langle \sigma, q \rangle \in \tau$ implies $\langle \sigma, q' \rangle \in \tau$ for any $q' \supseteq q$. So if G is a generic ideal, $p \in G$, and we can find $q \in G$ satisfying $\langle \sigma, q \rangle \in \tau$, then we can assume that q is an extension of p by replacing it with a common extension of p and q if necessary. Similarly, if q forces $\phi(\tau_1, \ldots, \tau_n)$ then so does any $q' \supseteq q$, simply because any generic ideal that contains q' must also contain q. So if G is a generic ideal, $p \in G$, and we can find $q \in G$ which forces $\phi(\tau_1, \ldots, \tau_n)$, then we can assume that q is an extension of p by replacing it with a common extension of p and q if necessary. In the sequel we will suppress this sort of argument and just say "by directedness".

Observe that the definition of the $\mathcal{F}_\alpha$ is absolute for $\mathbf{M}$. We say that the forcing relation for the formula $x_1 = x_2$ is *definable in* $\mathbf{M}$. This is all that really matters about the $\mathcal{F}_\alpha$; the precise formulation of conditions (i) and (ii) isn't crucially important.

The following simple fact about generic ideals is very useful.

Lemma 10.1. *Let G be a generic ideal of P, let $D \in \mathbf{M}$ be a subset of P, and suppose every element of G is compatible with some element of D. Then G must intersect D.*

Proof.   Let D' be the set of $p \in P$ which either extend some element of D or are incompatible with every element of D. Then D' belongs to $\mathbf{M}$ and is dense, so G must intersect D', and it follows from the hypothesis that some element of G extends some element of D. But this implies that G contains an element of D. $\square$

Corollary 10.2. *Let G be a generic ideal of P.*

(a) If $D \in \mathbf{M}$ is dense above some $p \in G$ then G must intersect D.
(b) If $G \subseteq A \in \mathbf{M}$ then there exists $p \in G$ such that every extension of p lies in A.

Proof.   (a) Fix $p \in G$. We know that every element of G is compatible with p, i.e., lies below an extension of p. So if D is dense above p then every element of G has an extension in D, and therefore every element of G is compatible with some element of D. Thus G must intersect D.

   (b) If $G \subseteq A$ then G does not intersect $P \setminus A$, and by the lemma there must be an element of G which is incompatible with every element of $P \setminus A$. So all of its extensions must lie in A. $\square$

Theorem 10.3. *Let τ_1 and τ_2 be P-names of name rank at most α.*

(a) A generic ideal G of P satisfies $\tau_1^G = \tau_2^G$ if and only if some $p \in G$ forces $\tau_1 = \tau_2$.
(b) An element $p \in P$ forces $\tau_1 = \tau_2$ if and only if $\langle p, \tau_1, \tau_2 \rangle \in \mathcal{F}_\alpha$.

Proof.   We prove the theorem by induction on α. Assume we know (a) and (b) for all $\beta < \alpha$. Nearly all of the technical content of the proof is contained in the following pair of claims (and the analogous assertions with the roles of τ_1 and τ_2 reversed).

   Define $A \subseteq P$ by letting $q_1 \in A$ if for each σ_1 with $\langle \sigma_1, q_1 \rangle \in \tau_1$ there exists $\langle \sigma_2, q_2 \rangle \in \tau_2$ such that $q_2 \supseteq q_1$ and $q_2 \Vdash \sigma_1 = \sigma_2$. For any generic ideal G of P we claim the following:

(1) If $\tau_1^G \subseteq \tau_2^G$ then $G \subseteq A$.
(2) If every extension of some $p \in G$ is in A then $\tau_1^G \subseteq \tau_2^G$.

   To prove (1), suppose $\tau_1^G \subseteq \tau_2^G$. Let $q_1 \in G$ and suppose $\langle \sigma_1, q_1 \rangle \in \tau_1$. Then $\sigma_1^G \in \tau_1^G \subseteq \tau_2^G$, so there must exist $\langle \sigma_2, q_2 \rangle \in \tau_2$ such that $q_2 \in G$ and $\sigma_1^G = \sigma_2^G$. By directedness we can assume $q_2 \supseteq q_1$. Also, by the induction hypothesis on part (a) of the theorem there exists $p \in G$ such that $p \Vdash \sigma_1 = \sigma_2$, so by directedness again we can assume $q_2 \Vdash \sigma_1 = \sigma_2$. This shows that $q_1 \in A$. We conclude that $G \subseteq A$.

To prove (2), suppose every extension of some $p \in G$ belongs to A. Now any element of τ_1^G has the form σ_1^G for some $\langle \sigma_1, q_1 \rangle \in \tau_1$ with $q_1 \in G$; by directedness, we can assume $q_1 \supseteq p$. Since every extension q_1' of q_1 now belongs to A and satisfies $\langle \sigma_1, q_1' \rangle \in \tau_1$, it follows that the set of q_2 which force $\sigma_1 = \sigma_2$ for some σ_2 with $\langle \sigma_2, q_2 \rangle \in \tau_2$ is dense above q_1. By the induction hypothesis on part (b) of the theorem this set belongs to $\mathbf{M}$. So by Corollary 10.2 (a) some such q_2 belongs to G, and this entails that $\sigma_1^G = \sigma_2^G \in \tau_2^G$. We conclude that $\tau_1^G \subseteq \tau_2^G$.

We proceed with the proof of Theorem 10.3. The reverse direction of part (a) is trivial. For the forward direction, let G be a generic ideal and suppose $\tau_1^G \subseteq \tau_2^G$. According to claim (1) we have $G \subseteq A$. We also know from the induction hypothesis on part (b) of the theorem that A lies in $\mathbf{M}$, so we can apply Corollary 10.2 (b) to find an element $p \in G$ all of whose extensions lie in A. Then claim (2) shows that p forces $\tau_1 \subseteq \tau_2$. A similar argument shows that if $\tau_2^G \subseteq \tau_1^G$ then there exists $p' \in G$ that forces $\tau_2 \subseteq \tau_1$. Thus, if $\tau_1^G = \tau_2^G$ then any common extension of p and p' in G will force $\tau_1 = \tau_2$. This completes the proof of the induction step for part (a).

For the forward direction of part (b), suppose that p forces $\tau_1 = \tau_2$. We must show that $\langle p, \tau_1, \tau_2 \rangle \in \mathcal{F}_\alpha$. Now condition (i) in the definition of $\mathcal{F}_\alpha$ says that every extension of p belongs to A. But for any $q_1 \supseteq p$ we can find a generic ideal G that contains q_1, and then claim (1) yields $q_1 \in A$. So every extension of p indeed belongs to A. A similar argument verifies condition (ii) in the definition of $\mathcal{F}_\alpha$.

For the reverse direction of part (b), suppose $\langle p, \tau_1, \tau_2 \rangle \in \mathcal{F}_\alpha$ and let G be a generic ideal that contains p; we must show that $\tau_1^G = \tau_2^G$. But again, condition (i) in the definition of $\mathcal{F}_\alpha$ says that every extension of p belongs to A, so claim (2) yields $\tau_1^G \subseteq \tau_2^G$. A similar argument shows that $\tau_2^G \subseteq \tau_1^G$. $\qquad\square$

Chapter 11

The Fundamental Theorem

Fix a forcing notion P. We will now generalize Theorem 10.3 to arbitrary formulas. The remaining cases are easier because, working in $\mathbf{M}$, we can define the forcing relation for the formula $x_1 \in x_2$ in terms of the forcing relation for $x_1 = x_2$, and the forcing relation for complex formulas can be constructed from the forcing relations for their constituent formulas. So we no longer need the induction on α employed in the definition of $\mathcal{F}_\alpha$ when the forcing relation for $x_1 = x_2$ was being defined in terms of itself.

Let ϕ be a formula in the language of set theory. In $\mathbf{M}$, we define a transfinite sequence of sets $(\mathcal{F}_\alpha^\phi)$ which describe the forcing relation for ϕ. If ϕ is an atomic equality this was done in Chapter 10; otherwise, let n be the number of unquantified variables in ϕ — since we can rename variables at will, there is no harm in assuming these are the variables $x_1, \ldots, x_n$ — and depending on whether ϕ has the form $x_1 \in x_2$, $\neg\psi$, $\psi_1 \to \psi_2$, or $(\forall x)\psi(x, x_1, \ldots, x_n)$ define $\mathcal{F}_\alpha^\phi$ to be the set of $(n{+}1)$-tuples $\langle p, \tau_1, \ldots, \tau_n \rangle \in P \times N_\alpha^n$ such that

($\in$) the set of q which force $\tau_1 = \sigma$ for some $\langle \sigma, q \rangle \in \tau_2$ is dense above p;

($\neg$) no extension of p forces $\psi(\tau_1, \ldots, \tau_n)$;

($\to$) every extension of p that forces $\psi_1(\tau_1, \ldots, \tau_n)$ has an extension that forces $\psi_2(\tau_1, \ldots, \tau_n)$;

($\forall$) for any P-name τ, the set of q which force $\psi(\tau, \tau_1, \ldots, \tau_n)$ is dense above p.

Each of these definitions can be framed in $\mathbf{M}$ because each refers to a forcing relation that we will be able to inductively assume is definable in $\mathbf{M}$.

For each ϕ the following statement is provable in ZFC^+.

Theorem Scheme 11.1. *(Fundamental theorem of forcing) Let $\tau_1, \ldots, \tau_n$ be P-names of name rank at most α.*

(a) For any generic ideal G of P we have $\mathbf{M}[G] \models \phi(\tau_1^G, \ldots, \tau_n^G)$ if and only if some $p \in G$ forces $\phi(\tau_1, \ldots, \tau_n)$.

(b) An element $p \in P$ forces $\phi(\tau_1, \ldots, \tau_n)$ if and only if $\langle p, \tau_1, \ldots, \tau_n \rangle \in \mathcal{F}_\alpha^\phi$.

Proof. The case where ϕ is an atomic equality was proven in Theorem 10.3. Otherwise, depending on whether ϕ has the form $x_1 \in x_2$, $\neg\psi$, $\psi_1 \to \psi_2$, or $(\forall x)\psi(x, x_1, \ldots, x_n)$ define A to be the set of $q \in P$ such that

$(\in)$ there exists $\langle \sigma, q' \rangle \in \tau_2$ such that $q' \supseteq q$ and q' forces $\tau_1 = \sigma$;

$(\neg)$ q does not force $\psi(\tau_1, \ldots, \tau_n)$;

$(\to)$ either q does not force $\psi_1(\tau_1, \ldots, \tau_n)$ or q has an extension which forces $\psi_2(\tau_1, \ldots, \tau_n)$;

$(\forall)$ for every P-name τ there is an extension of q which forces $\psi(\tau, \tau_1, \ldots, \tau_n)$.

Observe that in each case $\langle p, \tau_1, \ldots, \tau_n \rangle \in \mathcal{F}_\alpha^\phi$ if and only if every extension of p belongs to A. Now for any generic ideal G of P we claim the following:

(1) If $\mathbf{M}[G] \models \phi(\tau_1^G, \ldots, \tau_n^G)$ then $G \subseteq A$.

(2) If every extension of some $p \in G$ is in A then $\mathbf{M}[G] \models \phi(\tau_1^G, \ldots, \tau_n^G)$.

Proving the two claims for each possible form of ϕ is an amusing exercise. (At some points we need to invoke previously proven instances of Theorem Scheme 11.1.) The rest of the argument is the same as in the proof of Theorem 10.3, except that we now get $A \in \mathbf{M}$ not by an induction on α but by invoking previously proven instances of Theorem Scheme 11.1. $\square$

For the rest of this chapter, fix a generic ideal G of P. In Theorem 9.5 we showed that $\mathbf{M}[G]$ satisfies five of the Zermelo-Fraenkel axioms. We illustrate the power of the fundamental theorem by using it to verify the remaining axioms.

We start by showing that every instance of the separation scheme can be proven, in ZFC^+, to hold in $\mathbf{M}[G]$. For each formula ϕ in the language of set theory with unquantified variables among $u, x_1, \ldots, x_n$, the following statement is provable in ZFC^+.

Theorem Scheme 11.2. *For any set $\tau^G \in \mathbf{M}[G]$ and any parameters $\tau_1^G, \ldots, \tau_n^G \in \mathbf{M}[G]$, the set*

$$y = \left\{ u \in \tau^G \mid \mathbf{M}[G] \models \phi(u, \tau_1^G, \ldots, \tau_n^G) \right\}$$

belongs to $\mathbf{M}[G]$.

Proof. Let τ be a P-name and define

$$\pi = \big\{ \langle \sigma, p \rangle \in \tau \mid p \Vdash \phi(\sigma, \tau_1, \ldots, \tau_n) \big\}.$$

This is a P-name, the crucial point being that the relevant forcing relation is definable in $\mathbf{M}$ (Theorem Scheme 11.1 (b)).

We show that $\pi^G = y$. First, let σ^G be a typical element of π^G, with $\langle \sigma, p \rangle \in \tau$, $p \in G$, and $p \Vdash \phi(\sigma, \tau_1, \ldots, \tau_n)$. Then we immediately have $\sigma^G \in \tau^G$ and $\mathbf{M}[G] \models \phi(\sigma^G, \tau_1^G, \ldots, \tau_n^G)$, so $\sigma^G \in y$. Conversely, let $u \in \tau^G$ and suppose $\mathbf{M}[G] \models \phi(u, \tau_1^G, \ldots, \tau_n^G)$. Then $u = \sigma^G$ for some $\langle \sigma, p \rangle \in \tau$ with $p \in G$, and now the crucial point is that some element of G forces $\phi(\sigma, \tau_1, \ldots, \tau_n)$ (Theorem Scheme 11.1 (a)). By directedness we can assume p forces this, so that $\langle \sigma, p \rangle \in \pi$ and hence $u = \sigma^G \in \pi^G$. We conclude that $\pi^G = y$. $\qquad \square$

For any P-name τ, we introduce the temporary notation $\hat{\tau} = \big\{ \langle \sigma, p \rangle \mid \sigma \subseteq \tau$ is a P-name and $p \in P \big\}$.

Theorem 11.3. *For any* $\tau^G \in \mathbf{M}[G]$ *we have* $\mathcal{P}(\tau^G) \cap \mathbf{M}[G] = \hat{\tau}^G \in \mathbf{M}[G]$.

Proof. It is clear that every element of $\hat{\tau}$ evaluates to a subset of τ^G. Conversely, let $\pi^G \subseteq \tau^G$ be any subset of τ^G in $\mathbf{M}[G]$. Define the P-name $\pi' = \big\{ \langle \sigma', p \rangle \in \tau \mid p \Vdash \sigma' \in \pi \big\}$. Then $\pi' \subseteq \tau$, so that $\pi'^G \in \hat{\tau}^G$. We will complete the proof by showing that $\pi'^G = \pi^G$.

The inclusion $\pi'^G \subseteq \pi^G$ is easy. For the reverse inclusion, given any element σ^G of $\pi^G \subseteq \tau^G$ there must exist $\langle \sigma', p \rangle \in \tau$ such that $p \in G$ and $\sigma'^G = \sigma^G$. By directedness we can assume that p forces $\sigma' = \sigma \in \pi$; then $\langle \sigma', p \rangle \in \pi'$ and hence $\sigma^G = \sigma'^G \in \pi'^G$. $\qquad \square$

Our proof of the replacement scheme uses a version of the cumulative hierarchy in $\mathbf{M}[G]$. For every ordinal α in $\mathbf{M}$, define $\tilde{V}_\alpha = \big\{ \sigma^G \mid \sigma \in N_\alpha \big\}$. As usual, let $\tilde{V}_{<\alpha} = \bigcup_{\beta < \alpha} \tilde{V}_\beta$ and $N_{<\alpha} = \bigcup_{\beta < \alpha} N_\beta$.

Corollary 11.4. *We have* $\tilde{V}_\alpha = \mathcal{P}(\tilde{V}_{<\alpha}) \cap \mathbf{M}[G]$ *for every ordinal* $\alpha \in \mathbf{M}$.

Proof. For each α define $\tau_\alpha = \big\{ \langle \sigma, p \rangle \mid \sigma \in N_\alpha$ and $p \in P \big\}$. Letting $\tau_{<\alpha} = \bigcup_{\beta < \alpha} \tau_\beta$, it is straightforward to check that $\hat{\tau}_{<\alpha} = \tau_\alpha$. The corollary now follows from Theorem 11.3 since $\tilde{V}_\alpha = \tau_\alpha^G$ and $\tilde{V}_{<\alpha} = \tau_{<\alpha}^G$. $\qquad \square$

The corollary effectively says that $\tilde{V}_\alpha = V_\alpha^{\mathbf{M}[G]}$, the αth set in the cumulative hierarchy relativized to $\mathbf{M}[G]$, although we need to prove that $\mathbf{M}[G]$ models ZFC before we can talk about the cumulative hierarchy in $\mathbf{M}[G]$.

We can now verify the replacement scheme. Its proof is simplified by invoking the separation scheme which we have already proven. For each formula ϕ in the language of set theory with unquantified variables among $u, v, x_1, \ldots, x_n$, the following statement is provable in ZFC^+.

Theorem Scheme 11.5. *Let τ^G be a set in $\mathbf{M}[G]$ and let $\tau_1^G, \ldots, \tau_n^G$ be parameters. Suppose that for each $u \in \tau^G$ there is a unique $v \in \mathbf{M}[G]$ such that $\phi(u, v, \tau_1^G, \ldots, \tau_n^G)$ holds in $\mathbf{M}[G]$. Then the set formed by replacing each $u \in \tau^G$ with the corresponding v belongs to $\mathbf{M}[G]$.*

Proof. For each $\langle \sigma, p \rangle \in \tau$ let $\alpha_{\sigma,p}$ be the least ordinal in $\mathbf{M}$, if one exists, such that p forces $\phi(\sigma, \pi, \tau_1, \ldots, \tau_n)$ for some $\pi \in N_{\alpha_{\sigma,p}}$. By replacement in $\mathbf{M}$, these ordinals constitute a set that belongs to $\mathbf{M}$, so their supremum α also belongs to $\mathbf{M}$. By separation in $\mathbf{M}[G]$, it will suffice to check that the set we are trying to construct is contained in $\tilde{V}_\alpha$. That is, for each $u \in \tau^G$ we need the corresponding v to belong to $\tilde{V}_\alpha$, meaning that we need it to be the value of a P-name of name rank at most α. But we know $u = \sigma^G$ for some $\langle \sigma, p \rangle \in \tau$ with $p \in G$, and $v = \pi^G$ for some P-name π, and therefore some $q \in G$ forces $\phi(\sigma, \pi, \tau_1, \ldots, \tau_n)$. By directedness we can assume $\langle \sigma, q \rangle \in \tau$, and then we must have that $q \Vdash \phi(\sigma, \pi', \tau_1, \ldots, \tau_n)$ for some $\pi' \in N_\alpha$. By the uniqueness assumption it follows that $\pi^G = \pi'^G$, and we conclude that $\pi^G \in \tilde{V}_\alpha$, as desired. $\square$

Finally, we verify the axiom of choice.

Theorem 11.6. *The axiom of choice holds in $\mathbf{M}[G]$.*

Proof. Fix $\tau^G \in \mathbf{M}[G]$. In $\mathbf{M}$ we can well-order the P-name τ. Then
$$\left\{ \langle \text{op}(\langle \sigma, p \rangle^{\smile}, \sigma), p \rangle \mid \langle \sigma, p \rangle \in \tau \right\}$$
is a P-name that evaluates to a function f that maps a subset of τ onto τ^G. Working in $\mathbf{M}[G]$, which we can do because we know it satisfies the Zermelo-Fraenkel axioms other than choice, we can then embed τ^G into τ by mapping each $x \in \tau^G$ to the least element of $f^{-1}(x)$, and use this to induce a well-ordering of τ^G. This shows that every set can be well-ordered in $\mathbf{M}[G]$. The axiom of choice follows, because given any set of nonempty sets we can well-order their union and then select the least element from each of them. $\square$

We conclude that the following metatheorem is provable in PA.

Metatheorem 11.7. *ZFC^+ proves the statement "If G is a generic ideal of a forcing notion then $\mathbf{M}[G] \models \phi$" for every axiom ϕ of ZFC.*

Chapter 12

Forcing CH

Working in ZFC^+, we can now prove that every forcing notion has a generic ideal G (Theorem 8.3) and that each axiom of ZFC holds in the generic extension $\mathbf{M}[G]$ associated to G (Metatheorem 11.7). According to Metatheorem 7.4, all we have left to do to establish in PA that some statement ϕ is relatively consistent with ZFC — that is, that the consistency of ZFC implies the consistency of ZFC $+ \phi$ — is to produce a forcing notion for which we can prove in ZFC^+ that ϕ holds in $\mathbf{M}[G]$, for some generic ideal G. As our first example, we will prove in this chapter that the continuum hypothesis is relatively consistent with ZFC. This result was first proven by Gödel using very different methods, but given the machinery of forcing that is now at our disposal we can make a much quicker argument.

As we explained in Chapter 8, the usual interpretation of a forcing notion is that it represents a family of partial constructions of some desired object. Since we are currently interested in creating a bijection between $\mathcal{P}(\mathbb{N})$ and $\aleph_1$, one natural interpretation of "partial construction" could be a bijection between a finite subset of $\mathcal{P}(\mathbb{N})$ and a finite subset of $\aleph_1$. We may call such an object a *finite partial bijection*. Thus, define P_1 to be the set of $f \in \mathbf{M}$ such that $\mathbf{M} \models$ "f is a bijection between a finite subset of $\mathcal{P}(\mathbb{N})$ and a finite subset of $\aleph_1$". In other words, P_1 is the set of finite partial bijections between $\mathcal{P}(\mathbb{N})^{\mathbf{M}}$ and $\aleph_1^{\mathbf{M}}$. Since forcing notions have to belong to $\mathbf{M}$, we have defined P_1 by relativizing to $\mathbf{M}$ the concept of a finite partial bijection between $\mathcal{P}(\mathbb{N})$ and $\aleph_1$. However, when speaking informally, one usually takes the perspective of $\mathbf{M}$ and simply says, for example, that one is forcing with finite partial bijections between $\mathcal{P}(\mathbb{N})$ and $\aleph_1$.

Let G be a generic ideal of P_1. Since any two functions in G have a common extension, it follows that all the functions in G are compatible as functions, so that we can patch them together (set-theoretically, take their union) to obtain a single function $\tilde{f}$. Moreover, any union of a directed family of bijections is still a bijection, so $\tilde{f}$ must be a bijection between a subset of $\mathcal{P}(\mathbb{N})^{\mathbf{M}}$ and a subset of $\aleph_1^{\mathbf{M}}$. But for each $A \in \mathcal{P}(\mathbb{N})^{\mathbf{M}}$ the set of finite partial bijections whose domain includes A will be a dense set that lies in $\mathbf{M}$, so that G must contain such a function, and this shows that the domain of $\tilde{f}$ will be all of $\mathcal{P}(\mathbb{N})^{\mathbf{M}}$; by similar reasoning, its range will be all of $\aleph_1^{\mathbf{M}}$. So $\tilde{f}$ will be a bijection between $\mathcal{P}(\mathbb{N})^{\mathbf{M}}$ and $\aleph_1^{\mathbf{M}}$.

But this is not quite what we want: if $\mathbf{M}[G]$ is to satisfy the continuum hypothesis, then we need it to contain a bijection between $\mathcal{P}(\mathbb{N})^{\mathbf{M}[G]}$ and $\aleph_1^{\mathbf{M}[G]}$. This is the subtlety that we alluded to in Chapter 8. It is not enough to have, in $\mathbf{M}[G]$, a bijection between the sets that played the roles of $\mathcal{P}(\mathbb{N})$ and $\aleph_1$ in $\mathbf{M}$. We need a bijection between the sets that play these roles in $\mathbf{M}[G]$.

How can $\mathcal{P}(\mathbb{N})$ and $\aleph_1$ change? The property of being an ordinal is absolute for any transitive set, so the set that plays the role of $\aleph_1$ in $\mathbf{M}$ continues to be an ordinal in $\mathbf{M}[G]$. Indeed, we already know that $\mathbf{M}$ and $\mathbf{M}[G]$ have the same ordinals (Proposition 9.4). However, the property of being a cardinal is not absolute. Certainly, if the ordinal α is not a cardinal in $\mathbf{M}$ then there is, in $\mathbf{M}$, a bijection between it and a smaller ordinal, and this bijection will also belong to $\mathbf{M}[G]$, so α cannot be a cardinal in $\mathbf{M}[G]$ either. New cardinals cannot appear. But new bijections between ordinals might very well appear in $\mathbf{M}[G]$, and this means that cardinals in $\mathbf{M}$ might not continue to be cardinals in $\mathbf{M}[G]$. In particular, $\aleph_1^{\mathbf{M}}$ could become countable in $\mathbf{M}[G]$. An easy way to see that this can really happen is by forcing with finite partial bijections between $\mathbb{N}$ and $\aleph_1$; this introduces a bijection between $\mathbb{N}$, which is absolute, and $\aleph_1^{\mathbf{M}}$.

Turning to $\mathcal{P}(\mathbb{N})$, the property of being a set of natural numbers is absolute, so every set of natural numbers in $\mathbf{M}$ continues to be a set of natural numbers in $\mathbf{M}[G]$. That is, $\mathcal{P}(\mathbb{N})^{\mathbf{M}} \subseteq \mathcal{P}(\mathbb{N})^{\mathbf{M}[G]}$. However, new sets of natural numbers could appear in $\mathbf{M}[G]$. In fact, this does happen when we force with finite partial bijections between $\mathcal{P}(\mathbb{N})$ and $\aleph_1$. For instance, the bijection $\tilde{f} : \mathcal{P}(\mathbb{N})^{\mathbf{M}} \cong \aleph_1^{\mathbf{M}}$ we constructed above from G can be used to define a subset A of $\mathbb{N}$ by the prescription

$$n \in A \qquad \leftrightarrow \qquad \tilde{f}(n) < \omega,$$

and we claim that this set cannot belong to $\mathbf{M}$. (Recall that $n \in \mathbb{N}$ implies $n \subset \mathbb{N}$, so that $\tilde{f}(n)$ makes sense.) To see this, given any set $B \subseteq \mathbb{N}$ that appears in $\mathbf{M}$ let D_B be the set of $f \in P_1$ whose domain contains a natural number n such that either $f(n) \geq \omega$ and $n \in B$, or $f(n) < \omega$ and $n \notin B$.

Then D_B belongs to $\mathbf{M}$, and it is dense because we can extend any finite partial bijection by defining it to meet the stated condition at some natural number not already in its domain. So G must meet D_B. This implies that for some n we will have

$$n \in A \qquad \leftrightarrow \qquad \tilde{f}(n) < \omega \qquad \leftrightarrow \qquad n \notin B.$$

Since B was arbitrary, we conclude that the set A, which can be constructed in $\mathbf{M}[G]$, does not equal any subset of $\mathbb{N}$ that appears in $\mathbf{M}$.

Forcing with finite partial bijections introduces new sets of natural numbers, and this prevents us from deducing that the continuum hypothesis holds in $\mathbf{M}[G]$. We must use *countable partial bijections*, bijections between countable subsets of $\mathcal{P}(\mathbb{N})$ and $\aleph_1$. Thus let P_2 be the set of all $f \in \mathbf{M}$ such that $\mathbf{M} \models$ "f is a bijection between a countable subset of $\mathcal{P}(\mathbb{N})$ and a countable subset of $\aleph_1$". We stress that although $\mathcal{P}(\mathbb{N})^{\mathbf{M}}$ and $\aleph_1^{\mathbf{M}}$ are themselves countable, within $\mathbf{M}$ they appear to be uncountable.

The key property relevant to P_2 is the following.

Definition 12.1. A set P is *ω-closed* if the union of every infinite increasing sequence $p_0 \subseteq p_1 \subseteq p_2 \subseteq \cdots$ of elements of P belongs to P.

Working within $\mathbf{M}$, an infinite increasing sequence (f_n) in P_2 would be an infinite sequence of countable partial bijections such that each f_{n+1} extends f_n. The set-theoretic union of any such sequence is clearly another countable partial bijection. So $\mathbf{M} \models$ "P_2 is ω-closed".

This property of P_2 ensures that no new functions with domain $\mathbb{N}$ appear in $\mathbf{M}[G]$.

Lemma 12.2. *Let P be a forcing notion such that $\mathbf{M} \models$ "P is ω-closed", let G be a generic ideal of P, and let $X \in \mathbf{M}$. Then any function $f : \mathbb{N} \to X$ in $\mathbf{M}[G]$ is already in $\mathbf{M}$.*

Proof. Suppose $f : \mathbb{N} \to X$ belongs to $\mathbf{M}[G]$ and let τ be a P-name for f. Find $p \in G$ which forces the statement "τ is a function from $\check{\mathbb{N}}$ to $\check{X}$".

Let $T \subseteq P \times \mathbb{N} \times X$ be the set of triples $\langle q, n, x \rangle$ such that $q \Vdash \mathrm{op}(\check{n}, \check{x}) \in \tau$. By Theorem Scheme 11.1 (b) the set T belongs to $\mathbf{M}$. We claim that for each $n \in \mathbb{N}$ the set of $q \in P$ such that $\langle q, n, x \rangle \in T$ for some x is dense above p. To see this, let p' be an extension of p and let G' be a generic ideal that contains p'. Since $p' \supseteq p$ it follows that $\mathbf{M}[G'] \models$ "$\tau^{G'}$ is a function from $\mathbb{N}$ to X"; therefore $\langle n, x \rangle \in \tau^{G'}$ for some $x \in X$, and hence some $q \in G'$ forces $\mathrm{op}(\check{n}, \check{x}) \in \tau$. By directedness we can assume $q \supseteq p'$, so the claim is proven.

Now, working in $\mathbf{M}$, for any $p' \supseteq p$ we can use the claim to choose an infinite sequence (p_n) in P and a corresponding sequence (x_n) in X such that $p' \subseteq p_0 \subseteq p_1 \subseteq \cdots$ and each p_n forces $\mathrm{op}(\check{n}, \check{x}_n) \in \tau$. Since $\mathbf{M} \models$ "P is ω-closed", $p^* = \bigcup p_n \in P$; but then p^* forces $\mathrm{op}(\check{n}, \check{x}_n) \in \tau$ for every n, so that $\tau^{G'} = (x_n)$ for any generic ideal G' containing p^*. Letting S be the set of functions from $\mathbb{N}$ to X that lie in $\mathbf{M}$, we therefore have $p' \subseteq p^* \Vdash \tau \in \check{S}$. Since $p' \supseteq p$ was arbitrary, we have shown that the set of elements of P which force $\tau \in \check{S}$ is dense above p, so by Corollary 10.2 (a) G must contain such an element. Thus $f \in S$. $\square$

We can now show that CH is relatively consistent with ZFC.

Theorem 12.3. *Let P be the set of all $f \in \mathbf{M}$ such that $\mathbf{M} \models$ "f is a bijection between a countable subset of $\mathcal{P}(\mathbb{N})$ and a countable subset of $\aleph_1$" and let G be a generic ideal of P. Then the continuum hypothesis holds in $\mathbf{M}[G]$.*

Proof. According to the lemma, $\mathbf{M}[G]$ does not introduce any new functions with domain $\mathbb{N}$. Since subsets of $\mathbb{N}$ correspond to functions from $\mathbb{N}$ into $\{0, 1\}$, it follows that $\mathbf{M}[G]$ does not introduce any new subsets of $\mathbb{N}$, i.e., $\mathcal{P}(\mathbb{N})^{\mathbf{M}[G]} = \mathcal{P}(\mathbb{N})^{\mathbf{M}}$. Also, $\mathbf{M}[G]$ cannot introduce a function from $\mathbb{N}$ onto $\aleph_1^{\mathbf{M}}$, which means that $\aleph_1^{\mathbf{M}}$ must remain uncountable in $\mathbf{M}[G]$, i.e., $\aleph_1^{\mathbf{M}[G]} = \aleph_1^{\mathbf{M}}$.

By Proposition 9.4 we have $G \in \mathbf{M}[G]$. Thus the union of all the functions in G, which is a bijection $\tilde{f}$ between a subset of $\mathcal{P}(\mathbb{N})^{\mathbf{M}}$ and a subset of $\aleph_1^{\mathbf{M}}$, belongs to $\mathbf{M}[G]$. For each $A \in \mathcal{P}(\mathbb{N})^{\mathbf{M}}$ the set of functions in P whose domain contains A belongs to $\mathbf{M}$ and is dense, and for each $\alpha \in \aleph_1^{\mathbf{M}}$ the set of functions in P whose range contains α belongs to $\mathbf{M}$ and is dense. So $\tilde{f}$ must be a bijection between $\mathcal{P}(\mathbb{N})^{\mathbf{M}} = \mathcal{P}(\mathbb{N})^{\mathbf{M}[G]}$ and $\aleph_1^{\mathbf{M}} = \aleph_1^{\mathbf{M}[G]}$. Thus the continuum hypothesis holds in $\mathbf{M}[G]$. $\square$

By Metatheorem 7.4, Theorem 8.3, Metatheorem 11.7, and Theorem 12.3, the following is provable in PA.

Metatheorem 12.4. *If ZFC is consistent, then so is ZFC + CH.*

Chapter 13

Forcing ¬ CH

In this chapter we will prove the relative consistency of the negation of the continuum hypothesis. This is the theorem that the technique of forcing was originally developed by Cohen to prove. Together with the result of the last chapter, it shows that CH is independent of ZFC (assuming ZFC is consistent).

The forcing notion we need here is slightly different from the one we used in the last chapter. Forcing with partial bijections between $\mathcal{P}(\mathbb{N})$ and some cardinal κ greater than $\aleph_1$ will yield a bijection between $\mathcal{P}(\mathbb{N})^{\mathbf{M}}$ and κ in $\mathbf{M}[G]$, but this is unhelpful because if we already had a bijection between $\mathcal{P}(\mathbb{N})^{\mathbf{M}}$ and $\aleph_1^{\mathbf{M}}$ in $\mathbf{M}$ then the end result would be a bijection between κ and $\aleph_1^{\mathbf{M}}$ in $\mathbf{M}[G]$. We would merely have collapsed the cardinality of κ.

Instead of trying to directly introduce a bijection between $\mathcal{P}(\mathbb{N})$ and, say, $\aleph_2$, we will aim to introduce a family of $\aleph_2$ distinct subsets of $\mathbb{N}$. Such a family can be encoded as a single subset of $\mathbb{N} \times \aleph_2$, or equivalently a function from $\mathbb{N} \times \aleph_2$ into $\{0, 1\}$. Thus we work with the set of finite partial functions from $\mathbb{N} \times \aleph_2^{\mathbf{M}}$ into $\{0, 1\}$, that is, functions from finite subsets of $\mathbb{N} \times \aleph_2^{\mathbf{M}}$ into $\{0, 1\}$. From the perspective of $\mathbf{M}$ this set is not ω-closed, but it has a different structural property that will give us what we want.

Definition 13.1. A *join-antichain* in a set P is a family of pairwise incompatible elements. P is *c.c.c.* if every join-antichain in P is countable.

The abbreviation "c.c.c." stands for "countable chain condition", a misleading term for the property it describes, but one that is unfortunately standard. The importance of this condition here is that it allows us to approximate functions in $\mathbf{M}[G]$ with functions in $\mathbf{M}$, in the following sense.

Lemma 13.2. *Let P be a forcing notion such that $\mathbf{M} \models$ "P is c.c.c.", let G be a generic ideal of P, let $X, Y \in \mathbf{M}$, and let $f : X \to Y$ be a function* *in* $\mathbf{M}[G]$. *Then there is a function* $g : X \to \mathcal{P}(Y)$ *in* $\mathbf{M}$ *such that for each* $x \in X$ *we have* $f(x) \in g(x)$ *and* $\mathbf{M} \models$ *"$g(x)$ is countable".*

Proof. Let τ be a P-name for f and let $p \in G$ force "τ is a function from $\check{X}$ to $\check{Y}$". For $x \in X$ define

$$g(x) = \big\{ y \in Y \;\big|\; \text{some } q \supseteq p \text{ forces op}(\check{x}, \check{y}) \in \tau \big\}.$$

Then g belongs to $\mathbf{M}$, and we have $f(x) \in g(x)$ since some $q \in G$, and therefore some $q \supseteq p$, must force op$(\check{x}, \check{y}) \in \tau$ for $y = f(x)$. We must show that $\mathbf{M} \models$ "$g(x)$ is countable".

To see this, keep x fixed and, working in $\mathbf{M}$, for each $y \in g(x)$ find $q_y \supseteq p$ such that $q_y \Vdash$ op$(\check{x}, \check{y}) \in \tau$. Then the q_y are incompatible since they all lie above p, which forces τ to evaluate to a function, and they force different values of this function on x. Thus since "P is c.c.c." is true in $\mathbf{M}$, "$g(x)$ is countable" must also be true in $\mathbf{M}$. $\qquad\square$

As we remarked in Chapter 12, $\mathbf{M}$ and $\mathbf{M}[G]$ always have the same ordinals, and every cardinal in $\mathbf{M}[G]$ is a cardinal in $\mathbf{M}$, but the converse can fail. We say that a forcing notion P *preserves cardinals* if $\mathbf{M}$ and $\mathbf{M}[G]$ have the same cardinals, for every generic ideal G of P.

Theorem 13.3. *Let P be a forcing notion such that* $\mathbf{M} \models$ *"P is c.c.c." Then P preserves cardinals.*

Proof. Let G be a generic ideal of P and let α be an infinite ordinal in $\mathbf{M}$ such that $\mathbf{M}[G] \models$ "α is not a cardinal". Then there must be an infinite ordinal $\beta < \alpha$ and a surjection $f : \beta \to \alpha$ in $\mathbf{M}[G]$. With $g : \beta \to \mathcal{P}(\alpha)$ in $\mathbf{M}$ as in the lemma, we get

$$\mathbf{M} \models \operatorname{card}(\alpha) = \operatorname{card}\Big(\bigcup_{\gamma \in \beta} g(\gamma) \Big) \leq \aleph_0 \cdot \operatorname{card}(\beta) = \operatorname{card}(\beta),$$

so that α is not a cardinal in $\mathbf{M}$ either. We conclude that $\mathbf{M}$ and $\mathbf{M}[G]$ have the same cardinals. $\qquad\square$

We still have to show that the forcing notion we plan to use is c.c.c. according to $\mathbf{M}$. This will be done using the next lemma, which is a theorem of ordinary set theory and can be proven in ZFC.

Its proof uses the following easy fact. If we are given a graph with uncountably many vertices but such that each vertex has only countably many neighbors, then there is an uncountable set of vertices no two of which are neighbors. Indeed, a transfinite sequence of vertices $\big\{ v_\alpha \;\big|\; \alpha < \aleph_1 \big\}$ no two of which are neighbors can be constructed using a greedy algorithm:

at each stage α we will have already selected a countable family of vertices $\{v_\beta \mid \beta < \alpha\}$, each of which has only countably many neighbors, so by the uncountability of the vertex set we can choose a new vertex v_α which is not a neighbor of any of them.

Lemma 13.4. *(Δ-system lemma) Let X be an uncountable family of finite sets. Then there is an uncountable subset $Y \subseteq X$ and a set R such that $A \cap B = R$ for any distinct $A, B \in Y$.*

Proof. For each $k \in \mathbb{N}$ let $X_k = \{A \in X \mid \mathrm{card}(A) = k\}$. Then some X_k is uncountable; fix such a value of k. Now $\emptyset$ is contained in every $A \in X_k$, and any set with more than k elements is not contained in any $A \in X_k$, so there must be a maximum possible size of a set that is contained in uncountably many $A \in X_k$. Thus there must be some set R with $0 \leq \mathrm{card}(R) < k$ such that uncountably many $A \in X_k$ contain R, but for any $x \notin R$ only countably many $A \in X_k$ contain $R \cup \{x\}$. Fix such a set R and let $X'_k = \{A \in X_k \mid R \subseteq A\}$.

Now define a graph whose vertices are the elements of X'_k and which has an edge between A and B if $A \cap B \neq R$. For each $A \in X'_k$ the set $A \setminus R$ is finite, and we know that each $x \in A \setminus R$ is contained in only countably many $B \in X'_k$, so A can have only countably many neighbors in this graph. Thus by the graph-theoretic fact mentioned before the lemma, we can find an uncountable family of sets in X'_k, no two of which are neighbors, which just means that the intersection of any two of them equals R. □

The set Y is the Δ-*system* and the set R is its *root*. In the following proof we will actually only need an infinite Δ-system, not an uncountable one.

Theorem 13.5. *Let P be the set of all finite partial functions from $\mathbb{N} \times \aleph_2^{\mathbf{M}}$ into $\{0,1\}$ and let G be a generic ideal of P. Then the continuum hypothesis fails in $\mathbf{M}[G]$.*

Proof. Working in $\mathbf{M}$, we show that P is c.c.c. Let S be an uncountable subset of P and let $X = \{\mathrm{dom}(f) \mid f \in S\}$. Observe that X must be uncountable because any finite set supports only finitely many functions into $\{0,1\}$, and thus any countable family of finite sets can be the domains of only countably many functions into $\{0,1\}$. Now since the Δ-system lemma is provable in ZFC, it holds in $\mathbf{M}$, so we can apply it to X to find an infinite subset $Y \subseteq X$ and a finite set $R \subset \mathbb{N} \times \aleph_2$ such that $A \cap B = R$ for any distinct $A, B \in Y$. Form a corresponding infinite subset $T \subseteq S$

by selecting, for each $A \in Y$, one function in S whose domain is A. We then have $\text{dom}(f) \cap \text{dom}(g) = R$ for any distinct $f, g \in T$. But there are only finitely many functions from R into $\{0,1\}$, so there must exist some distinct $f, g \in T$ such that $f|_R = g|_R$. These two functions are therefore compatible, so we have shown that any uncountable subset of P contains compatible elements. Thus, working in $\mathbf{M}$, we have proven that P is c.c.c.

It follows from Theorem 13.3 that $\aleph_2^{\mathbf{M}[G]} = \aleph_2^{\mathbf{M}}$. By Proposition 9.4 we have $G \in \mathbf{M}[G]$. Thus the union of all the functions in G, which is a function $\tilde{f}$ from a subset of $\mathbb{N} \times \aleph_2^{\mathbf{M}}$ into $\{0,1\}$, belongs to $\mathbf{M}[G]$. Moreover, for each $\langle n, \alpha \rangle \in \mathbb{N} \times \aleph_2^{\mathbf{M}}$ the set of finite partial functions whose domain contains $\langle n, \alpha \rangle$ belongs to $\mathbf{M}$ and is dense, so the domain of $\tilde{f}$ is all of $\mathbb{N} \times \aleph_2^{\mathbf{M}}$.

For each ordinal $\alpha < \aleph_2$ define

$$A_\alpha = \left\{ n \in \mathbb{N} \mid \tilde{f}(n, \alpha) = 1 \right\}.$$

For any distinct $\alpha, \beta < \aleph_2$, the set $D_{\alpha,\beta}$ of all $f \in P$ for which there exists $n \in \mathbb{N}$ such that $\langle n, \alpha \rangle, \langle n, \beta \rangle \in \text{dom}(f)$ and $f(n, \alpha) \neq f(n, \beta)$ is dense and belongs to $\mathbf{M}$; since G is generic, it must intersect each $D_{\alpha,\beta}$, and this shows that the sets A_α are distinct. Since these sets can be constructed from $\tilde{f}$ in $\mathbf{M}[G]$, it follows that $\mathbf{M}[G] \models$ "there are $\aleph_2^{\mathbf{M}[G]}$ distinct subsets of $\mathbb{N}$". Thus the continuum hypothesis fails in $\mathbf{M}[G]$. $\square$

By Metatheorem 7.4, Theorem 8.3, Metatheorem 11.7, and Theorem 13.5, the following is provable in PA.

Metatheorem 13.6. *If ZFC is consistent, then so is ZFC $+ \neg$ CH.*

Note that Theorem 13.5 does not guarantee there are exactly $\aleph_2^{\mathbf{M}[G]}$ distinct subsets of $\mathbb{N}$ in $\mathbf{M}[G]$, only at least that many. Indeed, if $2^{\aleph_0} > \aleph_2$ in $\mathbf{M}$ then this will still be true in $\mathbf{M}[G]$ since cardinals do not change and we only add sets to $\mathcal{P}(\mathbb{N})$. Forcing exact values of $2^{\aleph_0}$ takes a little more work. We will not pursue this subject, although we will incidentally construct a model in which $2^{\aleph_0}$ is exactly $\aleph_2$ in Chapter 23.

It is worth noting that in Theorem 13.5 we could have used any cardinal $\kappa > \aleph_2$ in place of $\aleph_2$, and the result would be a model of ZFC in which $2^{\aleph_0} \geq \kappa$. So $2^{\aleph_0}$ can be arbitrarily large, in this sense. It is relatively consistent with ZFC that $2^{\aleph_0} \geq \aleph_\omega$, that $2^{\aleph_0} \geq \aleph_{\aleph_1}$, etc.

Chapter 14

Families of Entire Functions*

In chapters marked with an asterisk we will discuss, in ZFC, applications of set theory to problems that arise in mainstream mathematics. These chapters have some interdependence, but no unstarred chapter will depend on material from any starred chapter. Since we have just shown that the continuum hypothesis is independent of ZFC (assuming ZFC is consistent), the next few chapters will explore some consequences of CH.

The topic of the present chapter is a theorem due to Erdös which relates the continuum hypothesis to entire analytic functions in a surprising way. We require two standard facts from complex analysis: first, if the set of points at which two entire analytic functions agree has a cluster point, then they must be equal; second, if an infinite sequence of entire analytic functions converges uniformly on compact sets then the limit function is also entire analytic.

The problem solved by Erdös asks whether it is possible to find an uncountable family of entire analytic functions which collectively take only countably many values at each point of $\mathbb{C}$. Before we get to his solution we give three results which show what happens under slightly different hypotheses. None of these associated results involves substantial set theory.

Proposition 14.1. *There is a family of $2^{\aleph_0}$ distinct C^∞ functions on $\mathbb{R}$ which collectively take at most two values at each point.*

Proof. We require a nonzero C^∞ function $f : [0,1] \to \mathbb{R}$ such that f and all of its derivatives vanish at $t = 0$ and $t = 1$. For example, we could take the function defined by the formula
$$f(t) = e^{1/(t^2 - t)} = e^{-1/t} e^{-1/(1-t)}$$
for $t \in (0,1)$ and $f(0) = f(1) = 0$. An elementary calculation shows that it has the desired properties.

Now for each subset $A \subseteq \mathbb{Z}$ define $f_A : \mathbb{R} \to \mathbb{R}$ by setting

$$f_A(n + t) = \begin{cases} f(t) & \text{if } n \in A \\ 0 & \text{if } n \notin A \end{cases}$$

for $n \in \mathbb{Z}$ and $t \in [0, 1)$. This is a family of $2^{\aleph_0}$ distinct C^∞ functions, each of which takes one of the two values $f(t)$ or 0 at the point $n + t$, for any $n \in \mathbb{Z}$ and $t \in [0, 1)$.   □

The functions in Proposition 14.1 could be made to vanish at infinity, if one wished, by multiplying the entire family by, say, e^{-t^2}.

In contrast, analytic functions exhibit a global rigidity which leads to a very different conclusion.

Proposition 14.2. *Let $\mathcal{F}$ be a family of entire analytic functions such that for each $z \in \mathbb{C}$ the set of values $\{f(z) \mid f \in \mathcal{F}\}$ is finite. Then $\mathcal{F}$ is finite.*

Proof.   For each n let A_n be the set of points in $\mathbb{C}$ at which the family $\mathcal{F}$ takes exactly n values. Then some A_n must be uncountable, which implies that it must contain a cluster point. (If for each $z \in A_n$ the distance from z to $A_n \setminus \{z\}$ is nonzero then we can place an open disc of half this radius around each point of A_n and thereby produce a family of non-overlapping discs in the plane, which has to be countable.) So find $z \in A_n$ and $(z_k) \subset A_n \setminus \{z\}$ such that $z_k \to z$.

There are exactly n values $w_1, \ldots, w_n$ attained by the functions in $\mathcal{F}$ at the point z. Let r be the minimum distance between pairs of these n points and for each $1 \leq i \leq n$ fix a function $f_i \in \mathcal{F}$ such that $f_i(z) = w_i$. Since the f_i are continuous, we can find $\epsilon > 0$ such that

$$|z' - z| < \epsilon \qquad \to \qquad |f_i(z') - w_i| < r/2$$

for $1 \leq i \leq n$. Now the open discs of radius $r/2$ about the w_i are disjoint, so whenever $|z' - z| < \epsilon$ the values $f_i(z')$ must be distinct, and it follows that for each k with $|z_k - z| < \epsilon$ these are precisely the n values attained by functions in $\mathcal{F}$ at z_k. Thus, for any $f \in \mathcal{F}$ continuity implies that if $f(z) = w_i$ then $f(z_k) = f_i(z_k)$ for sufficiently large k. By the first fact about analytic functions cited above, we conclude that every function in $\mathcal{F}$ equals one of the f_i. So $\mathcal{F}$ is finite.   □

Complex polynomials are even more rigid. For any n distinct points $z_1, \ldots, z_n \in \mathbb{C}$, there are $2^{\aleph_0}$ distinct complex polynomials of degree n which vanish at each z_i, namely the polynomials

$$p_a(z) = a(z - z_1) \cdots (z - z_n)$$

with $a \in \mathbb{C} \setminus \{0\}$ arbitrary. So if p is any complex polynomial of degree n which takes the values $p(z_i) = w_i$ at the points $z_1, \ldots, z_n$, then $\{p + p_a \mid a \in \mathbb{C}\}$ will be a family of $2^{\aleph_0}$ distinct complex polynomials which take the same values at these points. However, these points at which all the polynomials $p + p_a$ take the same value are exceptional.

Proposition 14.3. *Let $\mathcal{F}$ be an uncountable family of complex polynomials. Then except for at most finitely many exceptional points $z \in \mathbb{C}$, the set of values $\{f(z) \mid f \in \mathcal{F}\}$ is uncountable.*

Proof. For each n let $\mathcal{F}_n$ be the set of polynomials in $\mathcal{F}$ of degree n. For some n the set $\mathcal{F}_n$ must be uncountable; fix such a value of n. We will prove that there are at most n points $z \in \mathbb{C}$ at which the set of values $\{p(z) \mid p \in \mathcal{F}_n\}$ is countable. To see this, observe first that there is no nonzero polynomial of degree at most n which vanishes at $n + 1$ distinct points, so if p and q are degree n polynomials which agree at $n + 1$ points then their difference must be the zero polynomial. This shows that a degree n polynomial is determined by its values on any $n + 1$ distinct points. Now suppose we could find $n + 1$ points z at each of which the set of values $\{p(z) \mid p \in \mathcal{F}_n\}$ is countable. Let A be this set of $n + 1$ points. Then distinct polynomials in $\mathcal{F}_n$ have distinct restrictions to A by our previous observation, but there are only countably many possible restrictions to A. This contradicts the premise that $\mathcal{F}_n$ is uncountable, so we conclude that the set of values $\{p(z) \mid p \in \mathcal{F}_n\}$ is uncountable except for at most n values of z. □

Now we turn to Erdös's problem about families of entire functions which take only countably many values at each point. The solution to this problem depends on the continuum hypothesis.

Theorem 14.4. *The continuum hypothesis is true if and only if there is an uncountable family $\mathcal{F}$ of entire analytic functions such that for each $z \in \mathbb{C}$ the set of values $\{f(z) \mid f \in \mathcal{F}\}$ is countable.*

Proof. ($\Leftarrow$) Suppose CH fails. Let $\{f_\alpha \mid \alpha < \aleph_1\}$ be a transfinite sequence of distinct entire functions; we will find a point in $\mathbb{C}$ at which the f_α take on uncountably many values. (If this can be done for any set of exactly $\aleph_1$ functions, it obviously can be done for sets of higher cardinalities.)

As we noted in the proof of Proposition 14.2, any uncountable subset of $\mathbb{C}$ must have a cluster point, so two distinct entire functions cannot take the same values at uncountably many points. Now for $\alpha, \beta < \aleph_1$ define

$$S_{\alpha,\beta} = \left\{ z \in \mathbb{C} \mid f_\alpha(z) = f_\beta(z) \right\}.$$

Then $\bigcup_{\alpha \neq \beta} S_{\alpha,\beta}$ is a union of $\aleph_1^2 = \aleph_1$ countable sets and hence has cardinality at most $\aleph_1$. Since we are assuming CH fails, it follows that there exists a point $z_0 \in \mathbb{C}$ that does not belong to any $S_{\alpha,\beta}$. But this means that the functions f_α all take distinct values at z_0, so the set of these values is uncountable.

($\Rightarrow$) Suppose CH holds and enumerate the complex plane as $\mathbb{C} = \left\{ z_\alpha \mid \alpha < \aleph_1 \right\}$. We will construct a transfinite sequence $\left\{ f_\beta \mid \beta < \aleph_1 \right\}$ of distinct entire functions such that $f_\beta(z_\alpha) \in \mathbb{Q} + i\mathbb{Q}$, i.e., both coordinates of $f_\beta(z_\alpha)$ are rational, for all $\beta > \alpha$. This implies that the set of values of the f_β at each z_α is countable, since for each α the set $\left\{ f_\beta(z_\alpha) \mid \beta \leq \alpha \right\}$ is trivially countable and the set $\left\{ f_\beta(z_\alpha) \mid \beta > \alpha \right\}$ is contained in the countable set $\mathbb{Q} + i\mathbb{Q}$.

We construct the f_β inductively. For finite n we can take $f_n = \prod_{i=0}^{n-1}(z - z_i)$. If $\omega \leq \beta < \aleph_1$ and we have constructed f_γ for all $\gamma < \beta$, then the set $\left\{ f_\gamma \mid \gamma < \beta \right\}$ is countable so it can be reordered as an infinite sequence (g_n). Similarly, reorder the countable set $\left\{ z_\alpha \mid \alpha < \beta \right\}$ as (w_n). We have to find an entire function f_β which satisfies

(i) $f_\beta(w_n) \in \mathbb{Q} + i\mathbb{Q}$ for all n
(ii) $f_\beta(w_n) \neq g_n(w_n)$ for all n.

Condition (i) ensures that $f_\beta(z_\alpha) \in \mathbb{Q} + i\mathbb{Q}$ for all $\alpha < \beta$, and condition (ii) ensures that $f_\beta \neq f_\gamma$ for all $\gamma < \beta$.

A function f_β with the desired properties can be chosen in the form

$$f_\beta(z) = \epsilon_0 + \sum_{n=1}^{\infty} \epsilon_n \prod_{i=0}^{n-1}(z - w_i).$$

If the ϵ_n are small enough that $\left| \epsilon_n \prod_0^{n-1}(z - w_i) \right| \leq 2^{-n}$ for $|z| \leq n$ then the sum will converge uniformly on compact sets, and hence it will define an entire function. This merely puts an upper bound on each ϵ_n but otherwise allows them to be arbitrary. We now sequentially choose each ϵ_n in a way that the nth partial sum for f_β satisfies conditions (i) and (ii) for that value of n. This suffices because all later terms in the sum vanish at w_n. Thus, given $\epsilon_0, \ldots, \epsilon_{n-1}$, evaluate the partial sum $\epsilon_0 + \sum_{k=0}^{n-1} \epsilon_k \prod_{i=0}^{k-1}(z - w_i)$ at the point w_n and then choose ϵ_n in such a way that adding the new term $\epsilon_n \prod_{i=1}^{n-1}(z - w_i)$ achieves a value at $z = w_n$ that lies in $\mathbb{Q} + i\mathbb{Q}$ and is distinct from $g_n(w_n)$. This completes the construction. $\square$

Chapter 15

Self-Homeomorphisms of $\beta\mathbb{N} \setminus \mathbb{N}$, I*

The *Stone-Čech compactification* $\beta\mathbb{N}$ of $\mathbb{N}$ is the maximal compact Hausdorff space into which $\mathbb{N}$ densely embeds. One way to construct it is to amalgamate all functions from $\mathbb{N}$ to $[0,1]$ into a single map from $\mathbb{N}$ into a power of $[0,1]$, and then take the closure of the image of $\mathbb{N}$ in this cube. The result is a compact Hausdorff space $\beta\mathbb{N}$ which contains a copy of $\mathbb{N}$ and such that any function from $\mathbb{N}$ into any compact Hausdorff space has a unique continuous extension to $\beta\mathbb{N}$.

The elements of $\mathbb{N}$ are topologically distinguished from the rest of $\beta\mathbb{N}$ by the fact that they are all isolated, so any self-homeomorphism of $\beta\mathbb{N}$ must permute $\mathbb{N}$, and since $\mathbb{N}$ is dense in $\beta\mathbb{N}$ this determines the entire map. Self-homeomorphisms of the *Stone-Čech remainder* $\beta\mathbb{N} \setminus \mathbb{N}$ are more subtle. First of all, it is sufficient to start with an *almost permutation*, a bijection between two cofinite subsets of $\mathbb{N}$. Since the inverse of any almost permutation is also an almost permutation, using the universal property of $\beta\mathbb{N}$ it is not hard to see that every almost permutation of $\mathbb{N}$ induces a self-homeomorphism of $\beta\mathbb{N} \setminus \mathbb{N}$. Whether all self-homeomorphisms of the Stone-Čech remainder arise in this way turns out to be independent of ZFC, assuming ZFC is consistent. In this chapter we will prove a theorem due to Rudin which states that under CH there are self-homeomorphisms of $\beta\mathbb{N} \setminus \mathbb{N}$ that do not arise from almost permutations of $\mathbb{N}$.

We start by making a reduction. Every self-homeomorphism of a totally disconnected compact Hausdorff space X gives rise to a distinct automorphism of the poset of clopen subsets of X ordered by inclusion, and every order-automorphism of the poset of clopen subsets arises in this way. This follows from the general theory of Stone duality, or it can be proven directly without much trouble. Now the poset of clopen subsets of $\beta\mathbb{N}$ can be identified with $\mathcal{P}(\mathbb{N})$ and the poset of clopen subsets of

$\beta \mathbb{N} \setminus \mathbb{N}$ can be identified with $\mathcal{P}(\mathbb{N})/\text{fin}$, the quotient of $\mathcal{P}(\mathbb{N})$ by the equivalence relation which makes A equivalent to B if their symmetric difference $A \triangle B = (A \setminus B) \cup (B \setminus A)$ is finite. Thus we must analyze the automorphisms of $\mathcal{P}(\mathbb{N})/\text{fin}$. Denoting the equivalence classes of A and B by $[A]$ and $[B]$, the order relation on $\mathcal{P}(\mathbb{N})/\text{fin}$ is defined by setting $[A] \leq [B]$ if all but finitely many elements of A belong to B.

The set-theoretic operations of union, intersection, and complementation in $\mathcal{P}(\mathbb{N})$ all descend to well-defined operations in the quotient structure. We will work with *Boolean subalgebras* of $\mathcal{P}(\mathbb{N})/\text{fin}$, subsets which are stable under these operations.

Finitely generated Boolean subalgebras of $\mathcal{P}(\mathbb{N})$ have a simple structure. Let $A_1, \ldots, A_k \subseteq \mathbb{N}$. Then any set of the form $A_1' \cap \cdots \cap A_k'$, where each A_i' equals either A_i or A_i^c, belongs to the Boolean subalgebra $\mathcal{A}$ of $\mathcal{P}(\mathbb{N})$ generated by the A_i. Some of these intersections could be empty, but the nonempty ones are the blocks of a partition of $\mathbb{N}$, and if we let $B_1, \ldots, B_r$ be these blocks then the sets in $\mathcal{A}$ are precisely those subsets of $\mathbb{N}$ which can be expressed as a union of some of the B_i. Thus $\mathcal{A}$ is order-isomorphic to the power set of an r-element set. The B_i are the *atoms* of $\mathcal{A}$.

Passing to $\mathcal{P}(\mathbb{N})/\text{fin}$ will annihilate any B_i whose cardinality is finite but it has no other effect. Thus $\tilde{\mathcal{A}} = \big\{ [A] \mid A \in \mathcal{A} \big\}$ is the Boolean subalgebra of $\mathcal{P}(\mathbb{N})/\text{fin}$ generated by $[A_1], \ldots, [A_k]$ and it is order-isomorphic to the power set of an s-element set where s is the number of atoms of $\mathcal{A}$ whose cardinality is infinite.

Evidently, two finite Boolean subalgebras $\tilde{\mathcal{A}}$ and $\tilde{\mathcal{A}}'$ of $\mathcal{P}(\mathbb{N})/\text{fin}$ are order-isomorphic to each other if and only if they contain the same number of elements. Any order-isomorphism between $\tilde{\mathcal{A}}$ and $\tilde{\mathcal{A}}'$ restricts to a bijection between the atoms in $\tilde{\mathcal{A}}$ and $\tilde{\mathcal{A}}'$, and since it must preserve finite suprema it is determined by this restriction. Conversely, any bijection between the atoms of $\tilde{\mathcal{A}}$ and $\tilde{\mathcal{A}}'$ clearly extends to an order-isomorphism.

Now let us consider what happens when a finite Boolean subalgebra is augmented by one additional element. Suppose $\mathcal{A}$ is a finite Boolean subalgebra of $\mathcal{P}(\mathbb{N})$ and $A \subseteq \mathbb{N}$. Then the structure of the Boolean subalgebra generated by $\mathcal{A}$ and A is completely determined by saying which atoms of $\mathcal{A}$ are contained in A and which are disjoint from A — equivalently, for which i we have $B_i \subseteq A$ and for which i we have $A \subseteq B_i^c$. In the generated Boolean subalgebra the atoms of $\mathcal{A}$ which are neither contained in nor disjoint from A each split into two atoms. Things work the same way in $\mathcal{P}(\mathbb{N})/\text{fin}$; all that matters is the order relation between the elements of the original Boolean subalgebra and the new element being adjoined to it.

This shows that an order-isomorphism ϕ between two finite Boolean sub-algebras $\tilde{\mathcal{A}}$ and $\tilde{\mathcal{A}}'$ of $\mathcal{P}(\mathbb{N})/\text{fin}$ can be extended to the Boolean subalgebra generated by $\tilde{\mathcal{A}}$ and an additional element x, provided only that we map x to an element $\phi(x)$ which is positioned relative to the elements of $\tilde{\mathcal{A}}'$ in the same way x is positioned relative to the corresponding elements of $\tilde{\mathcal{A}}$. Moreover, once $\phi(x)$ is specified the extension is unique. Since every finitely generated Boolean subalgebra of $\mathcal{P}(\mathbb{N})/\text{fin}$ is actually finite, it follows that every Boolean subalgebra of $\mathcal{P}(\mathbb{N})/\text{fin}$ is a directed union of finite Boolean subalgebras, and we can draw the following conclusion.

Lemma 15.1. *Let $\phi : \tilde{\mathcal{A}} \cong \tilde{\mathcal{A}}'$ be an order-isomorphism between two Boolean subalgebras of $\mathcal{P}(\mathbb{N})/\text{fin}$. Also let $x, y \in \mathcal{P}(\mathbb{N})/\text{fin}$ and suppose that ϕ takes the elements of $\tilde{\mathcal{A}}$ that are less than (respectively, greater than) x to the elements of $\tilde{\mathcal{A}}'$ that are less than (respectively, greater than) y. Then ϕ extends uniquely to an order-isomorphism that takes x to y between the Boolean subalgebra generated by $\tilde{\mathcal{A}}$ and x and the Boolean subalgebra generated by $\tilde{\mathcal{A}}'$ and y.*

The next lemma will be used to show that if $\tilde{\mathcal{A}}$ and $\tilde{\mathcal{A}}'$ are countable then a suitable target element y can always be found — in fact, at least two can be found. Write $A \subseteq_* B$ if all but finitely many elements of A are contained in B, and write $A \subset_* B$ if $A \subseteq_* B$ and $B \not\subseteq_* A$.

Lemma 15.2. *Let (A_i), (B_j), and (C_k) be infinite sequences of subsets of $\mathbb{N}$ such that*

$$A_1 \subseteq_* A_2 \subseteq_* A_3 \subseteq_* \cdots \subseteq_* B_3 \subseteq_* B_2 \subseteq_* B_1$$

and such that $B_j \not\subseteq_* A_i$, $B_j \not\subseteq_* C_k$, and $C_k \not\subseteq_* A_i$ for all i, j, k. Then there are two subsets Y and Y' of $\mathbb{N}$ with infinite symmetric difference such that $A_i \subset_* Y \subset_* B_j$ for all i and j and $C_k \not\subseteq_* Y \not\subseteq_* C_k$ for all k, and similarly for Y'.

Proof. We recursively construct two infinite sequences (S_n) and (T_n) such that $S_1 \subset S_2 \subset \cdots \subset T_2 \subset T_1$ and with the property that $S_n \triangle A_n$ and $T_n \triangle B_n$ are finite for all n, as follows. Start with $S_1 = A_1$ and $T_1 = A_1 \cup B_1$. Given S_{n-1} and T_{n-1}, let $S_n = S_{n-1} \cup (A_n \cap T_{n-1})$ together with one point of $T_{n-1} \setminus (S_{n-1} \cup A_1 \cup \cdots \cup A_n)$ and, for each $1 \le i \le n$ such that $A_n \subset_* C_i$, one point of $T_{n-1} \setminus (S_{n-1} \cup C_i)$. Then let $T_n = S_n \cup (B_n \cap T_{n-1})$ minus one point of $(B_1 \cap \cdots \cap B_n) \setminus S_n$ and, for each $1 \le i \le n$ such that $C_i \subset_* B_n$, one point of $(T_{n-1} \cap C_i) \setminus S_n$. After this construction is complete let $Y = \bigcup S_n$. Verifying that Y has the desired properties is straightforward.

In order to produce Y' we can run through the construction again, this time also adding one point of $T_{n-1} \setminus Y$ to S_n at each step. □

Thus, in Lemma 15.1, if $\tilde{A}$ is countable then we can enumerate the elements $\phi(a)$ with $a \leq x$ as $[A_i]$, the elements $\phi(b)$ with $x \leq b$ as $[B_j]$, and the elements $\phi(c)$ with $c \not\leq x \not\leq c$ as $[C_k]$ (possibly with repetitions, if there are only finitely many elements of any of these types). Then we can apply Lemma 15.2 to the sets $A_i' = A_1 \cup \cdots \cup A_i$, $B_j' = B_1 \cap \cdots \cap B_j$, and C_k to generate two distinct possible targets $y = [Y]$ and $y' = [Y']$ for x.

Theorem 15.3. *Assume CH. Then there exist self-homeomorphisms of* $\beta\mathbb{N} \setminus \mathbb{N}$ *that do not arise from almost permutations of* $\mathbb{N}$.

Proof. It will suffice to find an order-automorphism of $\mathcal{P}(\mathbb{N})/\text{fin}$ that does not arise from an almost permutation of $\mathbb{N}$. Enumerate the order-automorphisms of $\mathcal{P}(\mathbb{N})/\text{fin}$ which arise from almost permutations of $\mathbb{N}$ as $\{\psi_\alpha \mid 1 \leq \alpha < \aleph_1\}$ and enumerate the elements of $\mathcal{P}(\mathbb{N})/\text{fin}$ as $\{x_\alpha \mid 1 \leq \alpha < \aleph_1\}$. We will recursively construct countable Boolean subalgebras $\mathcal{A}_\alpha$ and $\mathcal{B}_\alpha$ of $\mathcal{P}(\mathbb{N})/\text{fin}$ and order-isomorphisms $\phi_\alpha : \mathcal{A}_\alpha \cong \mathcal{B}_\alpha$ such that

(i) if $\beta < \alpha$ then $\mathcal{A}_\beta \subset \mathcal{A}_\alpha$, $\mathcal{B}_\beta \subset \mathcal{B}_\alpha$, and $\phi_\alpha|_{\mathcal{A}_\beta} = \phi_\beta$
(ii) $\phi_\alpha \neq \psi_\alpha|_{\mathcal{A}_\alpha}$
(iii) x_α belongs to both $\mathcal{A}_\alpha$ and $\mathcal{B}_\alpha$.

Given this result, we define $\phi : \mathcal{P}(\mathbb{N})/\text{fin} \to \mathcal{P}(\mathbb{N})/\text{fin}$ by the condition $\phi|_{\mathcal{A}_\alpha} = \phi_\alpha$ to obtain an order-automorphism that differs from every ψ_α.

Let $\mathcal{A}_0 = \mathcal{B}_0 = \{[\emptyset], [\mathbb{N}]\}$ and let ϕ_0 be the identity map. At stage $\alpha \geq 1$, first let $\mathcal{A}_{<\alpha} = \bigcup_{\beta<\alpha} \mathcal{A}_\beta$. If $x_\alpha \notin \mathcal{A}_{<\alpha}$ then let $x_\alpha' = x_\alpha$; otherwise choose x_α' arbitrarily in the complement of $\mathcal{A}_{<\alpha}$. Let $\mathcal{A}_\alpha'$ be the Boolean subalgebra generated by $\mathcal{A}_{<\alpha}$ and x_α' and, by the comment preceding the theorem, extend $\phi_{<\alpha} = \bigcup_{\beta<\alpha} \phi_\beta$ to an order-isomorphism $\phi_\alpha' : \mathcal{A}_\alpha' \to \mathcal{B}_\alpha'$ where $\mathcal{B}_\alpha'$ is the Boolean subalgebra generated by $\mathcal{B}_{<\alpha} = \bigcup_{\beta<\alpha} \mathcal{B}_\beta$ and $\phi_\alpha'(x_\alpha')$. Since we have at least two candidates for the target element $\phi_\alpha'(x_\alpha')$, we can choose one that does not equal $\psi_\alpha(x_\alpha')$. To ensure that x_α belongs to $\mathcal{B}_\alpha$, let $\mathcal{B}_\alpha$ be the Boolean subalgebra generated by $\mathcal{B}_\alpha'$ and x_α and extend $(\phi_\alpha')^{-1}$ to an order-isomorphism ϕ_α^{-1} between $\mathcal{B}_\alpha$ and some Boolean subalgebra $\mathcal{A}_\alpha$ that contains $\mathcal{A}_\alpha'$. This completes the construction. □

Since for each α we can define ϕ_α in two different ways, with careful bookkeeping the proof of Theorem 15.3 can be used to show that under CH there are $2^{\aleph_1}$ distinct self-homeomorphisms of $\beta\mathbb{N} \setminus \mathbb{N}$, as opposed to only $\aleph_1$ self-homeomorphisms arising from almost permutations.

Chapter 16

Pure States on $\mathcal{B}(H)$*

In typical applications of the continuum hypothesis we construct something uncountable (an uncountable family of entire functions, an automorphism of $\mathcal{P}(\mathbb{N})/\text{fin}$) by means of a transfinite process which at each stage involves only countable data (a countable family of entire functions, an isomorphism between countable Boolean subalgebras of $\mathcal{P}(\mathbb{N})/\text{fin}$). In this chapter we will apply this technique to a question about pure states on $\mathcal{B}(H)$, the space of bounded linear operators from a separable infinite-dimensional complex Hilbert space H to itself.

$\mathcal{B}(H)$ is equipped with the *operator norm* defined by $\|A\| = \sup\{\|Av\| \mid v \in H, \|v\| = 1\}$, which makes it a nonseparable Banach space. We can also define the product of two operators to be their composition, and in addition every operator A has an adjoint operator A^* which is characterized by the formula $\langle Av, w \rangle = \langle v, A^*w \rangle$ for all $v, w \in H$. Thus it is natural to work with subsets of $\mathcal{B}(H)$ which are stable under these operations.

Definition 16.1. A *C*-algebra* is a closed linear subspace of $\mathcal{B}(H)$ that is stable under products and adjoints. It is *unital* if it contains the identity operator I. A *state* on a unital C*-algebra $\mathcal{A}$ is a bounded linear functional $\rho : \mathcal{A} \to \mathbb{C}$ such that $\|\rho\| = \rho(I) = 1$. It is *pure* if it cannot be expressed as $\rho = (\rho_1 + \rho_2)/2$ for two other states ρ_1 and ρ_2.

It follows from a general result in convexity theory, the Krein-Milman theorem, that pure states always exist. Pure states on C*-algebras are of interest because they give rise to irreducible representations. We will say a little more about this in Chaper 19.

The norm of a linear functional ρ on $\mathcal{A}$ is defined to be $\|\rho\| = \sup\{|\rho(A)| \mid A \in \mathcal{A}, \|A\| = 1\}$. Thus, if $\rho(I) = 1$ then $\|\rho\|$ must be at least

1. So in verifying that ρ is a state it is sufficient to check that $\|\rho\| \leq 1$.

For example, if v is a unit vector in H then the map $A \mapsto \langle Av, v \rangle$ is a state on $\mathcal{B}(H)$ because $\langle Iv, v \rangle = \langle v, v \rangle = 1$ and for any $A \in \mathcal{B}(H)$ the Cauchy-Schwarz inequality yields $|\langle Av, v \rangle| \leq \|Av\|\|v\| \leq \|A\|$. Moreover, it is pure, although showing this takes a little more work.

More generally, let (e_n) be an orthonormal basis of H and let $\xi \in \beta\mathbb{N}$. Since any bounded function from $\mathbb{N}$ to $\mathbb{C}$, i.e., any bounded complex infinite sequence (a_n), extends continuously to $\beta\mathbb{N}$, we can evaluate the extension at ξ; we denote this value $\lim_{n \to \xi} a_n$. Then the map

$$A \mapsto \lim_{n \to \xi} \langle Ae_n, e_n \rangle$$

is also a pure state on $\mathcal{B}(H)$. Verifying that it is a state is routine, but checking purity is surprisingly difficult. We say that this state is *diagonalized* by (e_n). The main result of this chapter, due to Akemann and the author, is that under CH there exist pure states on $\mathcal{B}(H)$ that are not diagonalizable, i.e., not diagonalized by any orthonormal basis. As of this writing it is open whether the assertion that all pure states on $\mathcal{B}(H)$ are diagonalizable is relatively consistent with ZFC.

The *rank* of an operator is the dimension of its range, so an operator is said to have finite rank if its range is finite-dimensional. An operator is *compact* if it can be approximated in norm by finite rank operators. The set of all compact operators on H is denoted $\mathcal{K}(H)$. We omit the proof of the following key C*-algebraic lemma.

Lemma 16.2. *Let $\mathcal{A} \subset \mathcal{B}(H)$ be a separable unital C*-algebra which contains $\mathcal{K}(H)$. Suppose ρ is a pure state on $\mathcal{A}$ which is zero on every compact operator and (e_n) is an orthonormal basis of H. Then there is a pure state $\tilde{\rho}$ on $\mathcal{B}(H)$ that extends ρ and an operator $P \in \mathcal{B}(H)$ such that $0 < \tilde{\rho}(P) < 1$ but $\langle Pe_n, e_n \rangle = 0$ or 1 for all n.*

The point is that $\lim_{n \to \xi} \langle Pe_n, e_n \rangle$ must be 0 or 1 for any $\xi \in \beta\mathbb{N}$, so the condition $0 < \tilde{\rho}(P) < 1$ implies that (e_n) cannot diagonalize $\tilde{\rho}$.

Now let ρ be a pure state on a C*-algebra $\mathcal{A}$ and let $A \in \mathcal{A}$. If whenever ρ_1 and ρ_2 are states such that $\rho = (\rho_1 + \rho_2)/2$ we must have $\rho_1(A) = \rho_2(A) = \rho(A)$, then we say that ρ is *pure on A*. Thus, ρ is pure if and only if it is pure on A for every $A \in \mathcal{A}$, and in fact, being pure on a dense subset of $\mathcal{A}$ is sufficient to ensure that ρ is pure.

Lemma 16.3. *Let ρ be a pure state on $\mathcal{B}(H)$ and let $S \subset \mathcal{B}(H)$ be a countable subset. Then there is a separable unital C*-algebra $\mathcal{A}$ that contains S such that the restriction of ρ to $\mathcal{A}$ is pure.*

Proof. To start with, note that the unital C*-algebra $\mathcal{A}$ generated by any countable set of operators S is separable. This is because the set of polynomials in the elements of S and their adjoints with complex rational coefficients is dense in $\mathcal{A}$.

Let $A \in \mathcal{B}(H)$. We claim that there is a countable set T which contains A and with the property that the restriction of ρ to any separable unital C*-algebra containing T is pure on A. To see this, suppose first that for each $n \in \mathbb{N}$ we can find a countable set T_n containing A with the property that whenever $\mathcal{A}$ is a separable unital C*-algebra containing T_n and ρ_1 and ρ_2 are states on $\mathcal{A}$ such that $\rho|_{\mathcal{A}} = (\rho_1 + \rho_2)/2$, we have $|\rho_1(A) - \rho(A)| < 1/n$. Then taking $T = \bigcup T_n$ would verify the claim. So if the claim fails there must exist $\epsilon > 0$ and a family $\{ \mathcal{A}_u \mid u \in X \}$ of separable unital C*-algebras, each one equipped with a pair of states ρ_1^u and ρ_2^u such that $\rho|_{\mathcal{A}_u} = (\rho_1^u + \rho_2^u)/2$ and $|\rho_1^u(A) - \rho(A)| \geq \epsilon$, and such that every countable set $T \subset \mathcal{B}(H)$ is contained in some $\mathcal{A}_u$. Since the $\mathcal{A}_u$'s are separable, this last condition implies that the family $\{ \mathcal{A}_u \mid u \in X \}$ is directed by inclusion, i.e., any two $\mathcal{A}_u$'s are contained in a third.

Now construct βX by amalgamating all the functions from X to $[0,1]$ into a single map from X into a power of $[0,1]$ and then taking the closure of the image of X in this cube. For each $u \in X$, let $C_u = \{ v \in X \mid \mathcal{A}_u \subseteq \mathcal{A}_v \}$. Since the $\mathcal{A}_u$ are directed by inclusion, the family $\{ C_u \mid u \in X \}$ has the finite intersection property. Thus, letting $\overline{C}_u$ be the closure of C_u in βX, we have $\bigcap_{u \in X} \overline{C}_u \neq \emptyset$ by compactness. Let ξ belong to this intersection and, for each $B \in \mathcal{B}(H)$, define $\rho_1(B) = \lim_{u \to \xi} \rho_1^u(B)$ and $\rho_2(B) = \lim_{u \to \xi} \rho_2^u(B)$. Then ρ_1 and ρ_2 are states on $\mathcal{B}(H)$ such that $\rho = (\rho_1 + \rho_2)/2$ and $|\rho_1(A) - \rho(A)| \geq \epsilon$, contradicting purity of ρ. This establishes the claim.

Now define an infinite sequence $(\mathcal{A}_k)$ of separable unital C*-algebras as follows. First let $\mathcal{A}_1$ be the unital C*-algebra generated by S. Given $\mathcal{A}_k$, let $\{ A_n \}$ be a countable dense subset of $\mathcal{A}_k$, and for each n find a countable set $T_n \subset \mathcal{B}(H)$ containing A_n such that the restriction of ρ to any separable unital C*-algebra containing T_n is pure on A_n. Then let $\mathcal{A}_{k+1}$ be the unital C*-algebra generated by $\bigcup T_n$. Finally, let $\mathcal{A}$ be the closure of $\bigcup \mathcal{A}_k$. This is a separable unital C*-algebra, and the restriction of ρ to $\mathcal{A}$ is pure on a dense subset of $\mathcal{A}$, and hence it is pure. $\square$

We need one more observation. Let α be a limit ordinal and suppose that for each $\beta < \alpha$ we have a unital C*-algebra $\mathcal{A}_\beta$ equipped with a pure state ρ_β, such that $\beta \leq \beta'$ implies $\mathcal{A}_\beta \subseteq \mathcal{A}_{\beta'}$ and $\rho_{\beta'}|_{\mathcal{A}_\beta} = \rho_\beta$. Then the closure $\mathcal{A}$ of $\bigcup_{\beta < \alpha} \mathcal{A}_\beta$ is a C*-algebra and $\bigcup_{\beta < \alpha} \rho_\beta$ extends by continuity to a state ρ on $\mathcal{A}$. Moreover, ρ is pure because if ρ'_1 and ρ'_2 are states on $\mathcal{A}$ such that $\rho = (\rho'_1 + \rho'_2)/2$, then for all $\beta < \alpha$ purity of ρ_β implies that $\rho'_1|_{\mathcal{A}_\beta} = \rho'_2|_{\mathcal{A}_\beta}$, and by continuity this implies that $\rho'_1 = \rho'_2$.

Theorem 16.4. *Assume CH. Then there is a pure state on $\mathcal{B}(H)$ that is not diagonalizable.*

Proof.   Enumerate the operators in $\mathcal{B}(H)$ as $\{A_\alpha \mid 1 \leq \alpha < \aleph_1\}$ and the orthonormal bases of H as $\{(e_n^\alpha) \mid 1 \leq \alpha < \aleph_1\}$. We recursively construct a transfinite sequence of separable unital C*-algebras $\mathcal{A}_\alpha \subset \mathcal{B}(H)$, $\alpha < \aleph_1$, together with pure states ρ_α on $\mathcal{A}_\alpha$, such that for all $\alpha \geq 1$

(i) $\beta \leq \alpha$ implies $\mathcal{A}_\beta \subseteq \mathcal{A}_\alpha$ and $\rho_\alpha|_{\mathcal{A}_\beta} = \rho_\beta$
(ii) $A_\alpha \in \mathcal{A}_\alpha$
(iii) $\mathcal{A}_\alpha$ contains an operator P_α satisfying $\langle P_\alpha e_n^\alpha, e_n^\alpha \rangle = 0$ or 1 for all n but such that $0 < \rho_\alpha(P_\alpha) < 1$.

At the initial stage we take $\mathcal{A}_0 = \mathcal{K}(H) + \mathbb{C} \cdot I$ and define ρ_0 by setting $\rho_0(A + zI) = z$. It is a routine exercise in C*-algebra to verify that $\mathcal{A}_0$ is a separable unital C*-algebra and ρ_0 is a pure state on $\mathcal{A}_0$. To carry out the construction at any subsequent stage $\alpha \geq 1$, first let $\mathcal{A}_{<\alpha}$ be the closure of $\bigcup_{\beta < \alpha} \mathcal{A}_\beta$ and, using the comment we made just before the theorem, let $\rho_{<\alpha}$ be the pure state on $\mathcal{A}_{<\alpha}$ whose restriction to each $\mathcal{A}_\beta$ equals ρ_β. Then use Lemma 16.2 to find a pure state $\tilde{\rho}$ on $\mathcal{B}(H)$ that extends $\rho_{<\alpha}$ and an operator $P_\alpha \in \mathcal{B}(H)$ such that $0 < \tilde{\rho}(P_\alpha) < 1$ but $\langle P_\alpha e_n^\alpha, e_n^\alpha \rangle = 0$ or 1 for all n. Since $\mathcal{A}_{<\alpha}$ is separable, Lemma 16.3 now provides us a separable unital C*-algebra $\mathcal{A}_\alpha$ which contains $\mathcal{A}_{<\alpha}$, A_α, and P_α, and such that the restriction ρ_α of $\tilde{\rho}$ to $\mathcal{A}_\alpha$ is pure. This completes the construction.

Finally, again using the comment we made just before the theorem, define a pure state ρ on $\mathcal{B}(H) = \bigcup_{\alpha < \aleph_1} \mathcal{A}_\alpha$ by the condition $\rho|_{\mathcal{A}_\alpha} = \rho_\alpha$. Since $0 < \rho(P_\alpha) < 1$ for all α we infer that no orthonormal basis of H can diagonalize ρ. $\square$

Chapter 17

The Diamond Principle

·

Build a tree by starting with one base vertex at level 0, putting $\aleph_0$ vertices above it at level 1, putting $\aleph_0$ vertices above each of those at level 2, and so on, through all finite levels. This is called the *complete $\aleph_0$-$\aleph_0$ tree*. A vertex at level n can be reached by making a series of choices — which successor of the base vertex to select at level 1, which successor of the level 1 vertex just chosen to select at level 2, etc. — and so it is uniquely identified by a function from $n = \{0, \ldots, n-1\}$ into $\mathbb{N}$. One vertex lies above another if the function corresponding to the former is an extension of the function corresponding to the latter. The infinite paths all the way up the tree correspond to functions from $\mathbb{N}$ to itself.

The *complete $\aleph_1$-$\aleph_1$ tree* is constructed in a similar way, with $\aleph_1$ successors to each vertex and levels of every countable height. For any countable ordinal α, we can identify the vertices at level α with the functions from α into $\aleph_1$. The paths all the way up the tree correspond to functions from $\aleph_1$ to itself.

The diamond principle has to do with the problem of trying to obstruct all paths up the complete $\aleph_1$-$\aleph_1$ tree by deleting a single vertex at each nonzero level. The analogous task for the complete $\aleph_0$-$\aleph_0$ tree is clearly impossible: no matter which vertices are removed, since each vertex has infinitely many successors it will be easy to create an unobstructed path that extends all the way up the tree. All we have to do is to start at the base vertex and work our way up, at each level choosing any successor of the current vertex other than the one forbidden vertex at the next level.

In the case of the complete $\aleph_1$-$\aleph_1$ tree it might seem like it should be even easier to avoid hitting a single forbidden vertex at each level. But a moment's thought shows that this intuition only applies to successor stages. There is also a possibility of landing on a bad vertex at some limit stage,

65

and while this chance may be slim at any particular limit level, there are $\aleph_1$ different limit levels to worry about. If one tries to work out a systematic method for avoiding forbidden vertices at future limit levels, one quickly realizes that it is not so obvious this can be done.

Now if the continuum hypothesis is false then one can indeed always find a path up the complete $\aleph_1$-$\aleph_1$ tree that avoids all the forbidden vertices, no matter how the latter were chosen. So it is relatively consistent with ZFC that there is no way to block all paths. Conversely, we will show that it is also relatively consistent that all paths can be blocked. In fact the diamond principle is a bit stronger: it says that one vertex can be removed from each level of the tree in such a way that every path is repeatedly blocked, in a manner that we will now explain.

Definition 17.1. A subset C of $\aleph_1$ is *closed* if the supremum of every countable subset of C belongs to C. It is *unbounded* if for every $\beta < \aleph_1$ there exists $\alpha \in C$ such that $\alpha > \beta$. A *club* subset is one which is both closed and unbounded. The *diamond principle* ($\Diamond$) is the assertion that one vertex can be chosen from each level of the complete $\aleph_1$-$\aleph_1$ tree in such a way that for every club subset C of $\aleph_1$, every path up the tree passes through one of the chosen vertices at a level belonging to C.

More formally, $\Diamond$ states that there is a transfinite sequence of functions $h_\alpha : \alpha \to \aleph_1$ for $\alpha < \aleph_1$ such that for any club subset $C \subseteq \aleph_1$ and any function $f : \aleph_1 \to \aleph_1$, there exists $\alpha \in C$ such that $f|_\alpha = h_\alpha$. The sequence (h_α) can be thought of as selecting one vertex from each level of the complete $\aleph_1$-$\aleph_1$ tree. We call it a *total vertex selection*.

For instance, for any $\beta < \aleph_1$ the set $\{\alpha \mid \beta < \alpha < \aleph_1\}$ is club, so diamond asserts that any path up the tree must hit a selected vertex at some level beyond β. Since this is true for arbitrary β, the path must actually hit selected vertices at an unbounded, and hence uncountable, set of levels.

A subset of $\aleph_1$ that intersects every club subset is called *stationary*. Thus, diamond can be reformulated as the assertion that for any path f the set $\{\alpha \mid f|_\alpha = h_\alpha\}$ is stationary. The observation we just made amounts to saying that every stationary subset of $\aleph_1$ is unbounded.

The forcing notion P we use to show that diamond is relatively consistent is described as follows (relativizing everything to $\mathbf{M}$). Each element of P is a sequence $p = \{h_\beta \mid \beta < \alpha\}$, where $\alpha < \aleph_1$ and each h_β is a function from β into $\aleph_1$. Thus, p chooses one vertex of the complete $\aleph_1$-$\aleph_1$ tree from each level below α. We call it a *partial vertex selection*. We call α the *height* of p.

We have $p \subseteq q$ if and only if the height of q is at least the height of p and every vertex selected by p is also selected by q. A generic ideal G of P may be thought of as making a generic choice of one vertex from each level of the complete $\aleph_1$-$\aleph_1$ tree. Now, working in $\mathbf{M}$, it is easy to see that if C is any unbounded subset of $\aleph_1$ and $f : \aleph_1 \to \aleph_1$ is any path up the tree, then the set of partial vertex selections that match f at some level in C is dense in P. Thus, the vertex sequence in $\mathbf{M}[G]$ determined by G must match f at a level in C. But of course this is not a proof that diamond holds in $\mathbf{M}[G]$, because we have to handle f and C in $\mathbf{M}[G]$, not just in $\mathbf{M}$.

Theorem 17.2. *Let P be the set of all $p \in \mathbf{M}$ such that $\mathbf{M} \models$ "p is a partial vertex selection from the complete $\aleph_1$-$\aleph_1$ tree" and let G be a generic ideal of P. Then the diamond principle holds in $\mathbf{M}[G]$.*

Proof. It is clear that $\mathbf{M} \models$ "P is ω-closed", so $\aleph_1^{\mathbf{M}} = \aleph_1^{\mathbf{M}[G]}$. The union of G is a total vertex selection $\{h_\alpha \mid \alpha < \aleph_1^{\mathbf{M}}\}$. We must show that this sequence verifies diamond in $\mathbf{M}[G]$. Let $\tilde{\Gamma}$ be a P-name for this sequence, as in Theorem 9.5.

Let f be a function in $\mathbf{M}[G]$ from $\aleph_1^{\mathbf{M}}$ to itself and let C be a club subset of $\aleph_1^{\mathbf{M}}$ that belongs to $\mathbf{M}[G]$. We have to prove that there exists $\alpha \in C$ such that $f|_\alpha = h_\alpha$. Let τ be a P-name for f, let σ be a P-name for C, and find $p \in G$ which forces "τ is a function from $\check{\aleph}_1^{\mathbf{M}}$ to itself and σ is a club subset of $\check{\aleph}_1^{\mathbf{M}}$".

We claim that for any $q \supseteq p$ of height α there exist $\beta > \alpha$, a function $g : \alpha \to \aleph_1^{\mathbf{M}}$ that belongs to $\mathbf{M}$, and an extension of q whose height is at least β and which forces both $\check{\beta} \in \sigma$ and $\tau|_{\check{\alpha}} = \check{g}$. To see this, let G' be a generic ideal that contains q and observe that since $p \Vdash$ "σ is unbounded", there exists an element of G' which forces $\check{\beta} \in \sigma$ for some $\beta > \alpha$. Also, letting g be the restriction of $\tau^{G'}$ to the countable ordinal α, we have $g \in \mathbf{M}$ by Lemma 12.2, so there exists an element of G' which forces $\tau|_{\check{\alpha}} = \check{g}$. By directedness, there must exist an element of G' which extends q and forces both conditions. By one more application of directedness, we can assume that the height of this element is at least β. This proves the claim.

Fix $q \supseteq p$. Working in $\mathbf{M}$, we construct a sequence $p_0 \subseteq p_1 \subseteq \cdots$ in P as follows. Start by letting $p_0 = q$. Given p_n, let its height be α_n and use the claim to find $\beta_n > \alpha_n$, a function $g_n : \alpha_n \to \aleph_1$, and an extension p_{n+1} of p_n whose height is at least β_n and which forces both $\check{\beta}_n \in \sigma$ and $\tau|_{\check{\alpha}_n} = \check{g}_n$. This shows how we obtain p_{n+1}.

Let $p^* = \bigcup p_n$. The height of p^* is $\alpha^* = \sup \alpha_n = \sup \beta_n$. For each n

we have $p^* \Vdash \check{\beta}_n \in \sigma$ since $p^* \supseteq p_n$, and since $p^* \supseteq p$ and $p \Vdash$ "σ is closed", it follows that $p^* \Vdash \check{\alpha}^* \in \sigma$. Let g^* be the union of the functions g_n, so that $p^* \Vdash \tau|_{\check{\alpha}^*} = \check{g}^*$, and let q' be the partial vertex selection of height $\alpha^* + 1$ which extends p^* by the one additional element $g^* : \alpha^* \to \aleph_1^{\mathbf{M}}$ at level α^*. Then any generic ideal G' that contains q' gives rise to a total vertex selection whose value at α^* equals g^* and hence equals the restriction of $\tau^{G'}$ to α^*. In other words, q' forces $\tilde{\Gamma}$ to match τ at some level in σ.

Since we have shown that any $q \supseteq p$ has an extension q' with this property, Corollary 10.2 (a) implies that G must contain such an element. We conclude that there exists $\alpha \in C$ such that $f|_\alpha = h_\alpha$. So in $\mathbf{M}[G]$ the set of α at which $f|_\alpha = h_\alpha$ is stationary.　　　　□

An odd feature of the preceding proof is that a claim is proven using generic ideals G' that do not belong to $\mathbf{M}$, but this claim is then used to perform a construction in $\mathbf{M}$. We did something similar earlier, in Lemma 12.2. Why is this legal?

The short answer is that all the reasoning in the proof is carried out in ZFC$^+$, and in the parts of the proof where we "work in $\mathbf{M}$" all this means is that those assertions and constructions are relativized to $\mathbf{M}$. "Working in $\mathbf{M}$" is feasible because we know that $\mathbf{M}$ satisfies the ZFC axioms, so that typical set-theoretic constructions, if relativized to $\mathbf{M}$ and given data that lies in $\mathbf{M}$, will produce a result that also lies in $\mathbf{M}$. There is nothing wrong with first working outside of $\mathbf{M}$ in order to prove that some set constructed in $\mathbf{M}$ is nonempty, and then, "in $\mathbf{M}$", performing a further construction involving this set that relies on the fact that it is nonempty.

On the other hand, since $\mathbf{M}$ is not assumed to have any special properties besides modelling ZFC, it might seem unlikely that this technique could give us information that could not be deduced within $\mathbf{M}$ using the ZFC axioms alone. In fact, the claim in the proof of Theorem 17.2 (and the claim in Lemma 12.2) could be proven working only in $\mathbf{M}$, by using properties of the forcing relation derived in Chapter 11. It is just a little easier to draw the conclusion using generic ideals.

By the usual argument, the following is provable in PA:

Metatheorem 17.3. *If ZFC is consistent, then so is ZFC* $+ \diamondsuit$.

As we mentioned earlier, the diamond principle implies the continuum hypothesis, so this incidentally yields a new proof of Metatheorem 12.4.

Chapter 18

Suslin's Problem, I*

The real line is dense (between any two points there is a third), unbounded (there is no least or greatest element), complete (every bounded set has a least upper bound and a greatest lower bound), and separable. Conversely, it is not too hard to show that any totally ordered set with these properties must be order-isomorphic to $\mathbb{R}$. *Suslin's problem* asks whether "separable" can be weakened to *topologically c.c.c.*, the condition that there is no uncountable collection of disjoint open intervals. A totally ordered set that is dense, unbounded, complete, and topologically c.c.c., but not separable, is called a *Suslin line*.

The existence of Suslin lines is independent of ZFC, assuming ZFC is consistent. In this chapter we will show that diamond implies Suslin lines exist, a result due to Jensen. We approach the problem using trees.

Definition 18.1. A *tree* is a partially ordered set $\mathcal{T}$ which contains a least element $0_{\mathcal{T}}$ and has the property that for every $x \in \mathcal{T}$ the set $x^< = \{y \in \mathcal{T} \mid y < x\}$ is well-ordered.

The *height* of $x \in \mathcal{T}$ is the ordinal isomorphic to $x^<$. A *level* of $\mathcal{T}$ is the set of all vertices of a given height. The *height* of $\mathcal{T}$ is the smallest ordinal that exceeds the heights of all of its elements.

A *branch* of $\mathcal{T}$ is a maximal totally ordered subset. An *antichain* is a set of pairwise incomparable elements.

A *normal Suslin tree* is a tree with the following properties:

(i) its height is $\aleph_1$
(ii) every antichain is countable
(iii) every vertex has infinitely many immediate successors
(iv) every vertex has extensions at all higher levels
(v) at limit levels, no two vertices have the same set of predecessors.

69

An easy observation is that a normal Suslin tree has no uncountable branches. If B is a branch of a normal Suslin tree, then for each vertex $v \in B$ we can choose an immediate successor of v that does not belong to B by property (iii); the resulting family of vertices will be an antichain, and hence it must be countable by property (ii), showing that B itself must have been countable.

Lemma 18.2. *If normal Suslin trees exist then Suslin lines exist.*

Proof. Suppose we are given a normal Suslin tree. Each vertex v has exactly $\aleph_0$ immediate successors; let $\prec_v$ be a total ordering on the set of immediate successors of v which makes it order-isomorphic to $\mathbb{Q}$. Then let $\mathcal{L}$ be the set of branches of $\mathcal{T}$. Given two distinct branches B_1 and B_2, by property (v) of normal Suslin trees the first level at which they differ must be a successor level. So there is a vertex v which belongs to both B_1 and B_2, but such that $v_1 \in B_1$ and $v_2 \in B_2$ for two distinct successors v_1 and v_2 of v. Set $B_1 \prec B_2$ if and only if $v_1 \prec_v v_2$. This totally orders $\mathcal{L}$.

It is easy to see that $\mathcal{L}$ is dense and has no least or greatest element. To see that it is topologically c.c.c., let $\{I_u\}$ be a family of disjoint open intervals in $\mathcal{L}$. For each u find $B_1, B_2 \in I_u$ such that $B_1 \prec B_2$, let v be the highest vertex belonging to both B_1 and B_2, and let $v_1 \in B_1$ and $v_2 \in B_2$ be its immediate successors with $v_1 \prec_v v_2$. Then find a successor w_u of v which lies between v_1 and v_2. Now every branch that passes through w_u belongs to I_u, and this implies that the vertices w_u are incomparable, i.e., they constitute an antichain. Since every antichain in $\mathcal{T}$ is countable, it follows that the family $\{I_u\}$ must be countable. So $\mathcal{L}$ is topologically c.c.c.

Next, let S be any countable family of branches of $\mathcal{T}$. As we observed just before the lemma, every branch of $\mathcal{T}$ is countable, so there is a level α such that each branch in S terminates below α. Let v be a vertex at level α and find distinct branches B_1 and B_2 which pass through v. Then no branch in S can lie between B_1 and B_2. This shows that the set S is not dense in $\mathcal{L}$. So we conclude that $\mathcal{L}$ is not separable.

We have seen that $\mathcal{L}$ has all the properties of a Suslin line except completeness. Completing it via Dedekind cuts yields a Suslin line. □

We are going to construct a normal Suslin tree, using diamond to prevent uncountable antichains from developing. The key fact, given in the following lemma, will tell us that any maximal antichain in the tree we construct must restrict to a maximal antichain in the truncations of the tree at a club set of levels.

Lemma 18.3. *Let T be a tree of height $\aleph_1$, let A be a maximal antichain in T, and for each $\alpha < \aleph_1$ let T_α be the subtree of T consisting of all vertices of height less than α. Suppose that each T_α is countable. Then*

$$C = \left\{ \alpha \mid A \cap T_\alpha \text{ is a maximal antichain in } T_\alpha \right\}$$

is a club subset of $\aleph_1$.

Proof. $A \cap T_\alpha$ is an antichain in T_α for any α. Suppose $\alpha_n \in C$ for each $n \in \mathbb{N}$ and let $\alpha = \sup\{\alpha_n\}$; then any element of T_α belongs to some T_{α_n} and hence is comparable to some element of $A \cap T_{\alpha_n}$. Thus every element of T_α is comparable to some element of $A \cap T_\alpha$, which means that $A \cap T_\alpha$ is a maximal antichain in T_α. So C is closed.

To see that C is unbounded, let $\alpha_1 < \aleph_1$. Then T_{α_1} is countable and every element of T_{α_1} is comparable to some element of A, so there exists $\alpha_2 \geq \alpha_1$ such that every element of T_{α_1} is comparable to some element of $A \cap T_{\alpha_2}$. Then find $\alpha_3 \geq \alpha_2$ such that every element of T_{α_2} is comparable to some element of $A \cap T_{\alpha_3}$, and so on. Setting $\alpha = \sup\{\alpha_n\}$, we have that every element of T_α is comparable to some element of $A \cap T_\alpha$, i.e., $\alpha \in C$. Since $\alpha \geq \alpha_1$ and α_1 was arbitrary, this shows that C is unbounded. $\square$

Theorem 18.4. *Assume $\Diamond$. Then Suslin lines exist.*

Proof. By Lemma 18.2 it will suffice to show that a normal Suslin tree exists. Using diamond, let $\{h_\alpha \mid \alpha < \aleph_1\}$ be a sequence of functions $h_\alpha : \alpha \to \aleph_1$ such that for any function $f : \aleph_1 \to \aleph_1$ the set $\{\alpha \mid f|_\alpha = h_\alpha\}$ is stationary. For each α let $A_\alpha = h_\alpha^{-1}(1) \subseteq \alpha$. Then by considering the characteristic function of any subset $A \subseteq \aleph_1$ we can see that $\{\alpha \mid A_\alpha = A \cap \alpha\}$ is stationary.

We will construct a normal Suslin tree T whose vertices are precisely the countable ordinals. We take the ordinals in order, meaning that every time we add a vertex it will be the first ordinal not yet used.

As we construct the tree we will ensure that for each $\alpha < \aleph_1$ the subtree T_α consisting of all vertices whose height is less than α will satisfy the following two conditions:

(1) T_α will be countable

(2) for any $\beta_1 < \beta_2 < \alpha$, every vertex of height β_1 will lie below a vertex of height β_2.

This last condition ensures that if α is a limit ordinal then every vertex v in T_α belongs to branch in T_α whose height is α. To see this, let the height of

v be α_1, use the fact that α is countable to find a sequence $\alpha_1 < \alpha_2 < \cdots$ whose supremum is α, and recursively choose a sequence of vertices v_n of height α_n such that $v_1 = v$ and v_{n+1} lies above v_n for all n. Then there will be a branch of $\mathcal{T}_\alpha$ that passes through each v_n.

The construction of $\mathcal{T}$ goes as follows. We start by letting 0 be the base vertex. This single vertex constitutes the subtree $\mathcal{T}_1$. Having constructed $\mathcal{T}_{\alpha+1}$ for some successor ordinal $\alpha+1$, we construct level $\alpha+1$ of the tree by adding $\aleph_0$ immediate successors to each vertex at level α. By the induction hypothesis there were only countably many vertices in $\mathcal{T}_{\alpha+1}$, so we will be adding only countably many new vertices, and hence condition (1) is preserved. Similarly, since by hypothesis every vertex of height less than α lies below a vertex of height α, and we are adding successors to every vertex of height α, condition (2) is also preserved.

Now suppose we have constructed $\mathcal{T}_\alpha$ for some limit ordinal α. If the set A_α defined earlier is not a maximal antichain in $\mathcal{T}_\alpha$, then for each $v \in \mathcal{T}_\alpha$ we choose a branch that contains v and has height α, and we add one vertex at level α to lie above each of these branches. Since $\mathcal{T}_\alpha$ is countable we are adding only countably many new vertices, so condition (1) is preserved, and it is clear from the construction that condition (2) is also preserved. On the other hand, if A_α is a maximal antichain in $\mathcal{T}_\alpha$, then every $v \in \mathcal{T}_\alpha$ lies either above or below some element of A_α, so we can perform the same construction as in the first case, but this time making sure that the branch of height α that contains v also contains an element of A_α, for each v. This completes the description of the construction of $\mathcal{T}$.

We must verify that $\mathcal{T}$ is a normal Suslin tree. All the desired properties are immediate except property (ii), which states that every antichain is countable. To see this, let $A \subset \mathcal{T}$ be a maximal antichain. By Lemma 18.3 the set C of levels for which $A \cap \mathcal{T}_\alpha$ is a maximal antichain in $\mathcal{T}_\alpha$ is club. Moreover, the set of $\alpha \in C$ for which $\mathcal{T}_\alpha = \alpha$ (as sets) is club. (Closure is easy. For unboundedness, observe that $\alpha \subseteq \mathcal{T}_\alpha$ holds for all α, and we get equality at the supremum of any sequence (α_n) with the property that $\mathcal{T}_{\alpha_n} \subseteq \alpha_{n+1}$ for all n.) So it follows from diamond that there exists α such that $A \cap \mathcal{T}_\alpha$ is a maximal antichain in $\mathcal{T}_\alpha$ and $A_\alpha = A \cap \mathcal{T}_\alpha$. Then the construction of the αth level of $\mathcal{T}$ ensures that every vertex at level α lies above some element of $A \cap \mathcal{T}_\alpha$. But this implies that every vertex of height greater than α also lies above some element of $A \cap \mathcal{T}_\alpha$. So $A \cap \mathcal{T}_\alpha$ must not only be a maximal antichain in $\mathcal{T}_\alpha$, it must be a maximal antichain in $\mathcal{T}$. We conclude that $A = A \cap \mathcal{T}_\alpha$, and therefore A is countable. This shows that every antichain in $\mathcal{T}$ is countable. $\qquad\square$

Chapter 19

Naimark's Problem*

In Chapter 16 we defined C*-algebras to be subsets of $\mathcal{B}(H)$ with certain stability properties. It can also be fruitful to consider a C*-algebra abstractly, as an object which has the structure of a Banach algebra equipped with an involution and which can be realized, up to a bijection preserving all relevant structure, as a closed subspace of some $\mathcal{B}(H)$ that is stable under products and adjoints. (We now allow H to be nonseparable.) From this point of view, the same C*-algebra can be represented on different Hilbert spaces, in the following sense.

Definition 19.1. A *representation* of a C*-algebra $\mathcal{A}$ on a Hilbert space K is a bounded linear map $\pi : \mathcal{A} \to \mathcal{B}(K)$ which respects products and adjoints. A representation is *irreducible* if K cannot be decomposed into an orthogonal direct sum of nonzero subspaces, $K = K_1 \oplus K_2$, in such a way that each summand is invariant for the action of $\mathcal{A}$ (that is, $\pi(x)K_i \subseteq K_i$ for all $x \in \mathcal{A}$ and $i = 1, 2$). If $\pi' : \mathcal{A} \to \mathcal{B}(K')$ is another representation of $\mathcal{A}$, we say that π and π' are *unitarily equivalent* if there is an isometric isomorphism $U : K \cong K'$ such that $\pi(x) = U^{-1}\pi'(x)U$ for all $x \in \mathcal{A}$.

An isometric isomorphism U between two Hilbert spaces is also called a *unitary* operator. Such operators are characterized by the algebraic condition $U^*U = UU^* = I$.

Irreducible representations are the building blocks of representation theory. They are closely related to pure states: if $\pi : \mathcal{A} \to \mathcal{B}(K)$ is an irreducible representation of a unital C*-algebra $\mathcal{A}$ and v is a unit vector in K, then the map $\rho : x \mapsto \langle \pi(x)v, v \rangle$ is a pure state on $\mathcal{A}$, and conversely, every pure state is realized in this way for some irreducible representation. Moreover, if π and π' are irreducible representations and ρ and ρ' are pure states arising from π and π' in the preceding manner, then π and π' are unitarily equivalent if and only if there is a unitary $u \in \mathcal{A}$ such that $\rho(x) = \rho'(u^*xu)$ for all $x \in \mathcal{A}$. We say that ρ and ρ' are *unitarily equivalent* pure states.

The basic problem in representation theory is to classify all irreducible representations of a given C*-algebra up to unitary equivalence. For example, an early result of Naimark states that the algebra $\mathcal{K}(H)$ of compact operators on a Hilbert space H has only one irreducible representation up to unitary equivalence. *Naimark's problem* asks whether these are the only C*-algebras with this property. Akemann and the author showed that diamond implies they are not. As of this writing it is not known whether a positive answer to Naimark's question is relatively consistent with ZFC.

We require the following basic C*-algebra facts. If $(\mathcal{A}_\alpha)$ is a (possibly transfinite) sequence of C*-algebras which is increasing, i.e., $\beta \le \alpha$ implies $\mathcal{A}_\beta \subseteq \mathcal{A}_\alpha$, then the completion of $\bigcup \mathcal{A}_\alpha$ is also a C*-algebra; and if each $\mathcal{A}_\alpha$ is *simple* — has no nontrivial closed two-sided ideals — then this completion is also simple. We also present one C*-algebraic lemma without proof.

Lemma 19.2. *Let $\mathcal{A}$ be a simple, separable, unital C*-algebra and let ρ and ρ' be pure states on $\mathcal{A}$. Then there is a simple, separable, unital C*-algebra $\mathcal{B}$ which contains $\mathcal{A}$ and has the same unit, such that ρ and ρ' have unique extensions to (necessarily pure) states on $\mathcal{B}$, and these extensions are unitarily equivalent.*

The idea of Lemma 19.2 is to augment $\mathcal{A}$ with a unitary that relates ρ and ρ'. This is accomplished using a "crossed product" construction. The key point is that any two pure states on a simple separable C*-algebra are related by an automorphism of the algebra.

The next lemma suggests how diamond may be relevant to the construction of a C*-algebra all of whose pure states are equivalent. Specifically, it gives a clue as to how we are going to use diamond to control all the pure states on a C*-algebra which is constructed in $\aleph_1$ stages. If $(\mathcal{A}_\alpha)$ is an increasing transfinite sequence of C*-algebras, for each α let $\mathcal{A}_{<\alpha}$ be the completion of $\bigcup_{\beta<\alpha} \mathcal{A}_\beta$.

Lemma 19.3. *Let $(\mathcal{A}_\alpha)$, $\alpha < \aleph_1$, be an increasing transfinite sequence of separable unital C*-algebras. Let $\mathcal{A} = \bigcup \mathcal{A}_\alpha$ and let ρ be a pure state on $\mathcal{A}$. Then*

$$C = \left\{ \alpha \mid \rho|_{\mathcal{A}_{<\alpha}} \text{ is a pure state on } \mathcal{A}_{<\alpha} \right\}$$

is a club subset of $\aleph_1$.

Proof. Note that $\mathcal{A}$ is automatically complete because any convergent sequence (x_n) in $\mathcal{A}$ belongs to some $\mathcal{A}_\alpha$ and hence has a limit in $\mathcal{A}_\alpha$ which also belongs to $\mathcal{A}$. (Each x_n belongs to some $\mathcal{A}_{\alpha_n}$, and then we can take $\alpha = \sup \alpha_n$.)

The set C is unbounded by an argument similar to the one used to prove Lemma 16.3: starting with any $\mathcal{A}_\beta$, using separability we can find $\beta_2 \geq \beta_1 = \beta$ such that $\rho|_{\mathcal{A}_{\beta_2}}$ is pure on a dense set of $x \in \mathcal{A}_{\beta_1}$. Then find $\beta_3 \geq \beta_2$ such that $\rho|_{\mathcal{A}_{\beta_3}}$ is pure on a dense set of $x \in \mathcal{A}_{\beta_2}$, and so on. We get that $\beta^* = \sup \beta_n \geq \beta$ and $\rho|_{\mathcal{A}_{<\beta^*}}$ is pure. The comment just before Theorem 16.4 shows that C is closed. So C is a club subset of $\aleph_1$. $\quad\square$

We will build a counterexample to Naimark's problem by taking the union of an increasing transfinite sequence $(\mathcal{A}_\alpha)$ of simple, separable, unital C*-algebras. Since it is known that there are no separable counterexamples, we expect that each $\mathcal{A}_\alpha$ will have pairs of inequivalent pure states. Letting ρ and ρ' be one such pair, we can use Lemma 19.2 to construct the next algebra $\mathcal{A}_{\alpha+1}$ in such a way that ρ and ρ' each have unique extensions to $\mathcal{A}_{\alpha+1}$ and these extensions are equivalent.

This construction will achieve the goal of making one pair of pure states equivalent, but there may still be many other inequivalent pairs, and even worse, new pure states could appear in $\mathcal{A}_{\alpha+1}$. Each pure state on $\mathcal{A}_\alpha$ has at least one, but possibly many, extensions to pure states on $\mathcal{A}_{\alpha+1}$, and in addition new pure states can appear on $\mathcal{A}_{\alpha+1}$ whose restrictions to $\mathcal{A}_\alpha$ are not pure. So it might seem hard to believe that we can eventually catch all the pure states — but in the same way it is hard to believe we can block all the paths up a complete $\aleph_1$-$\aleph_1$ tree. What we can do is to accumulate a pool of equivalent pure states and, at each stage, use diamond to choose a new pure state to add to the pool. Lemma 19.3 can then be used to tell us that any pure state on the final algebra will have been captured at some stage of the construction.

In order to get the construction going we need to start with a simple, separable, infinite-dimensional, unital C*-algebra. There are many examples of these. For instance, for each $n \in \mathbb{N}$ unitally embed the $2^n \times 2^n$ complex matrix algebra $M_{2^n}(\mathbb{C}) \cong \mathcal{B}(\mathbb{C}^{2^n})$ in $M_{2^{n+1}}(\mathbb{C})$ by the map $A \mapsto \begin{bmatrix} A & 0 \\ 0 & A \end{bmatrix}$, and then take the completion of the union $\bigcup M_{2^n}(\mathbb{C})$. The resulting algebra will be simple since it is the completion of the union of an increasing sequence of simple C*-algebras, and it is separable, infinite-dimensional, and unital by inspection.

Theorem 19.4. *Assume* $\diamondsuit$. *Then there is a C*-algebra that has only one irreducible representation up to unitary equivalence but is not isomorphic to the algebra of compact operators on any Hilbert space.*

Proof. Let $\{h_\alpha : \alpha \to \aleph_1 \mid \alpha < \aleph_1\}$ be a diamond sequence. For $\alpha < \aleph_1$ we will recursively construct (1) a simple separable, unital C*-algebra $\mathcal{A}_\alpha$; (2) a distinguished pure state ρ_α on $\mathcal{A}_\alpha$; and (3) an injective map ϕ_α from the set of states on $\mathcal{A}_\alpha$ into $\aleph_1$. We will ensure that $\mathcal{A}_\beta \subseteq \mathcal{A}_\alpha$ (with the same unit) and that ρ_α is the unique extension of ρ_β to a state on $\mathcal{A}_\alpha$, for any $\beta \leq \alpha$.

Begin by letting $\mathcal{A}_0$ be any simple, separable, unital C*-algebra, ρ_0 any pure state on $\mathcal{A}_0$, and ϕ_0 any injective map from the set of states on $\mathcal{A}_0$ into $\aleph_1$. Such a map must exist since diamond implies the continuum hypothesis, so that any separable C*-algebra has at most $\aleph_1$ states.

Now let $\alpha > 0$ and suppose the construction has been completed for all $\beta < \alpha$. Let $\rho_{<\alpha}$ be the pure state on $\mathcal{A}_{<\alpha}$ whose restriction to $\mathcal{A}_\beta$ is ρ_β, for all $\beta < \alpha$. If there is a pure state ρ' on $\mathcal{A}_{<\alpha}$ that satisfies $\phi_\beta(\rho'|_{\mathcal{A}_\beta}) = h_\alpha(\beta)$ for all $\beta < \alpha$, then let $\mathcal{A}_\alpha$ be the C*-algebra given by Lemma 19.2 with $\rho = \rho_{<\alpha}$, and let ρ_α be the unique extension of $\rho_{<\alpha}$ to $\mathcal{A}_\alpha$. Otherwise, let $\mathcal{A}_\alpha = \mathcal{A}_{<\alpha}$ and $\rho_\alpha = \rho_{<\alpha}$. In either case, let ϕ_α be any injective map from the set of states on $\mathcal{A}_\alpha$ into $\aleph_1$. This completes the construction.

Let $\mathcal{A} = \bigcup_{\alpha < \aleph_1} \mathcal{A}_\alpha$ and let ρ be the pure state on $\mathcal{A}$ whose restriction to each $\mathcal{A}_\alpha$ equals ρ_α. We must show that every pure state on $\mathcal{A}$ is unitarily equivalent to ρ. Let ρ' be any pure state on $\mathcal{A}$; we will show that $\rho'|_{\mathcal{A}_\alpha}$ is unitarily equivalent to ρ_α for some α. This is sufficient because ρ is the unique state extension of ρ_α to $\mathcal{A}$, so conjugation by the unitary that makes ρ_α and $\rho'|_{\mathcal{A}_\alpha}$ equivalent shows that ρ' is the unique state extension of $\rho'|_{\mathcal{A}_\alpha}$ to $\mathcal{A}$, and it is unitarily equivalent to ρ.

Define $f : \aleph_1 \to \aleph_1$ by setting $f(\alpha) = \phi_\alpha(\rho'|_{\mathcal{A}_{<\alpha}})$. According to Lemma 19.3, there is a club set C of ordinals α at which $\rho'|_{\mathcal{A}_{<\alpha}}$ is pure, so by diamond there exists $\alpha < \aleph_1$ such that $\rho'|_{\mathcal{A}_{<\alpha}}$ is pure and $\phi_\beta(\rho'|_{\mathcal{A}_\beta}) = h_\alpha(\beta)$ for all $\beta < \alpha$. Then the construction of $\mathcal{A}_\alpha$ guarantees that $\rho'|_{\mathcal{A}_\alpha}$ is the unique state extension of $\rho'|_{\mathcal{A}_{<\alpha}}$ to $\mathcal{A}_\alpha$ and it is unitarily equivalent to ρ_α. We conclude that all pure states on $\mathcal{A}$ are equivalent.

Finally, $\mathcal{A}$ is not isomorphic to the algebra of compact operators on any Hilbert space because it is infinite-dimensional and unital. $\qquad\square$

Chapter 20

A Stronger Diamond

The diamond principle can be strengthened in various ways. We will prove the relative consistency of a version in which paths are blocked using vertices in a stationary set of levels. Recall that a subset of $\aleph_1$ is stationary if it intersects every club subset of $\aleph_1$.

Definition 20.1. For any stationary set $S \subseteq \aleph_1$, $\diamondsuit_S$ is the assertion that there is a sequence $\{h_\alpha \mid \alpha \in S\}$ of functions $h_\alpha : \alpha \to \aleph_1$ such that for any function $f : \aleph_1 \to \aleph_1$ the set $\{\alpha \in S \mid f|_\alpha = h_\alpha\}$ is stationary.

Thus, $\diamondsuit_S$ says that paths up the complete $\aleph_1$-$\aleph_1$ tree can be blocked by removing vertices only at levels belonging to S. We will denote by $(\forall S)\diamondsuit_S$ the assertion that $\diamondsuit_S$ holds for all stationary $S \subseteq \aleph_1$.

It is not so hard to adapt the proof of Theorem 17.2 to get $\diamondsuit_S$ in $\mathbf{M}[G]$ for every stationary S in $\mathbf{M}$:

Lemma 20.2. *Let $S \in \mathbf{M}$ be a subset of $\aleph_1^{\mathbf{M}}$ such that $\mathbf{M} \models$ "S is stationary" and let P be the set of $p \in \mathbf{M}$ such that $\mathbf{M} \models$ "p is a partial vertex selection from the complete $\aleph_1$-$\aleph_1$ tree". Then $\mathbf{M}[G] \models \diamondsuit_S$, for any generic ideal G of P.*

Proof. Fix a generic ideal G of P and let $\tilde{\Gamma}^G$ be the total vertex selection formed by taking the union of G. Suppose that according to $\mathbf{M}[G]$, τ^G is a path up the complete $\aleph_1$-$\aleph_1$ tree and σ^G is a club subset of $\aleph_1$, and find $p \in G$ which forces both of these outcomes. As in Theorem 17.2, $\mathbf{M} \models$ "for any $q \supseteq p$ of height α there exists an extension q_0 of q whose height is $\alpha_0 + 1 > \alpha$ and which forces $\tilde{\Gamma}$ to match τ at level $\check{\alpha}_0 \in \sigma$".

Work in $\mathbf{M}$. Iterating the preceding fact yields a transfinite increasing sequence of elements $q_\beta \in P$ which force matches on a club sequence (α_β). Since S is stationary in $\mathbf{M}$, we must have $\alpha_\beta \in S$ for some β, so for this

77

value of β the element $q_\beta \supseteq q$ forces $\tilde{\Gamma}$ to match τ at a level belonging to $\sigma \cap \check{S}$. Since $q \supseteq p$ was arbitrary, Corollary 10.2 (a) implies that some element of G must also force this. Thus, in $\mathbf{M}[G]$ the diamond sequence $\tilde{\Gamma}^G$ matches τ^G at some level in $\sigma^G \cap S$.   □

The preceding argument only works for $S \in \mathbf{M}$. The way we force $(\forall S)\Diamond_S$ is by augmenting $\mathbf{M}$ with many diamond sequences, enough that for any stationary set in $\mathbf{M}[G]$ there is a diamond sequence that is added to $\mathbf{M}[G]$ "independently" of it. Then we can break up the forcing construction into two steps in a way that allows us to treat the stationary set as belonging to the ground model when the independent diamond sequence is added.

The technical tool we need is *product forcing*. The basic result says that forcing with a product $P_1 \cdot P_2$ is equivalent to first forcing with P_1 to obtain a new model $\mathbf{M}[G_1]$, and then treating $\mathbf{M}[G_1]$ as the ground model and forcing with P_2. It is convenient to assume that P_1 and P_2 both contain $\emptyset$ (they are *rooted*) and that $p \cap q = \emptyset$ for all $p \in P_1$ and $q \in P_2$ (they are *independent*).

Definition 20.3. The *product* of two independent rooted forcing notions P_1 and P_2 is the set

$$P_1 \cdot P_2 = \left\{ p \cup q \mid p \in P_1 \text{ and } q \in P_2 \right\}.$$

We define $G_1 \cdot G_2$ similarly for any ideals $G_1 \subseteq P_1$ and $G_2 \subseteq P_2$.

Theorem 20.4. *Let P_1 and P_2 be independent rooted forcing notions and suppose G is a generic ideal of $P_1 \cdot P_2$. Then $G = G_1 \cdot G_2$ where G_1 is a generic ideal of P_1 relative to $\mathbf{M}$ and G_2 is a generic ideal of P_2 relative to $\mathbf{M}[G_1]$. We have $\mathbf{M}[G] = \mathbf{M}[G_1][G_2]$.*

Proof. Let $G_1 = G \cap P_1$ and $G_2 = G \cap P_2$. If $p \cup q \in G$ for $p \in P_1$ and $q \in P_2$ then $p = p \cup \emptyset$ and $q = \emptyset \cup q$ both belong to G and therefore $p \in G_1$ and $q \in G_2$, so that $p \cup q \in G_1 \cdot G_2$; conversely, if $p \cup q \in G_1 \cdot G_2$ for $p \in G_1$ and $q \in G_2$ then $p, q \in G$ and therefore $p \cup q \in G$. So $G = G_1 \cdot G_2$. It is easy to see that G_1 and G_2 are ideals.

To prove that G_1 is generic relative to $\mathbf{M}$, let $D_1 \in \mathbf{M}$ be a dense subset of P_1. Then $\left\{ p \cup q \mid p \in D_1 \text{ and } q \in P_2 \right\}$ is a dense subset of $P_1 \cdot P_2$ that lies in $\mathbf{M}$, so G must intersect this set, and thus G_1 must intersect D_1.

To prove that G_2 is generic relative to $\mathbf{M}[G_1]$, let $D_2 \in \mathbf{M}[G_1]$ be a dense subset of P_2, let $\tau \in \mathbf{M}$ be a P_1-name for D_2, and fix $p_0 \in G_1$ such that $p_0 \Vdash$ "τ is a dense subset of $\check{P}_2$". Define

$$D = \left\{ p \cup q \in P_1 \cdot P_2 \mid p \supseteq p_0 \text{ and } p \Vdash \check{q} \in \tau \right\} \in \mathbf{M};$$

we claim that D is dense above $p_0 = p_0 \cup \emptyset \in G$. To see this, let $p \cup q \supseteq p_0$. Then $p \Vdash$ "there exists $x \in \tau$ such that $x \supseteq \check{q}$" since $p \supseteq p_0$. So there exist $p' \supseteq p$ and $q' \supseteq q$ such that $p' \Vdash \check{q}' \in \tau$, and then $p' \cup q' \in D$ and $p' \cup q' \supseteq p \cup q$. This proves the claim.

Thus G must intersect D. But if $p \cup q \in G \cap D$ then $p \in G_1$, so $p \Vdash \check{q} \in \tau$ implies that $q \in \tau^{G_1} = D_2$. Thus G_2 intersects D_2, as desired.

For the final statement of the theorem, observe that $G_1, G_2 \in \mathbf{M}[G]$ and $G \in \mathbf{M}[G_1][G_2]$. Given this, the two inclusions $\mathbf{M}[G] \subseteq \mathbf{M}[G_1][G_2]$ and $\mathbf{M}[G_1][G_2] \subseteq \mathbf{M}[G]$ are both proven by a straightforward induction on name rank, working in $\mathbf{M}[G_1][G_2]$ and in $\mathbf{M}[G]$. $\square$

Next, we need a version of the Δ-system lemma (Lemma 13.4) for countable sets. This is provable in ZFC. Here we introduce the notation κ^+ for the next cardinal after κ; thus, if $2^{\aleph_0} = \aleph_\alpha$ then $(2^{\aleph_0})^+ = \aleph_{\alpha+1}$.

Lemma 20.5. *Let X be a family of more than $2^{\aleph_0}$ distinct countable sets. Then there is a subset $Y \subseteq X$ whose cardinality is greater than $2^{\aleph_0}$ and a set R such that $A \cap B = R$ for any distinct $A, B \in Y$.*

Proof. We claim that there exists a set Z of cardinality at most $2^{\aleph_0}$ such that for any set Z' of cardinality at most $2^{\aleph_0}$, some $A \in X$ is disjoint from $Z' \setminus Z$ and not contained in Z. Suppose otherwise. Then we can recursively construct a sequence $\{Z_\beta \mid \beta < \aleph_1\}$ by, for each β, finding $Z_\beta\ (= Z')$ which falsifies the claim for $Z_{<\beta} = \bigcup_{\gamma<\beta} Z_\gamma\ (= Z)$. So for each β, every $A \in X$ either intersects $Z_\beta \setminus Z_{<\beta}$ or is contained in $Z_{<\beta}$. Then $\bigcup_{\beta<\aleph_1} Z_\beta$ has cardinality at most $\aleph_1 \cdot 2^{\aleph_0} = 2^{\aleph_0}$, so it has at most $2^{\aleph_0}$ countable subsets and hence cannot contain every $A \in X$. But if some $A \in X$ is not contained in $\bigcup_{\beta<\aleph_1} Z_\beta$, then it would have to intersect $Z_\beta \setminus Z_{<\beta}$ for every β, which is impossible since it is countable. This proves the claim.

Now fix Z verifying the claim. Construct a sequence of sets $A_\alpha \in X$ for $\alpha < (2^{\aleph_0})^+$ by, for each α, finding $A_\alpha \in X$ which is disjoint from $(\bigcup_{\gamma<\alpha} A_\gamma) \setminus Z$ and not contained in Z. This produces $(2^{\aleph_0})^+$ distinct sets any two of which can only overlap within Z. Since there are at most $2^{\aleph_0}$ countable subsets of Z, we can find $(2^{\aleph_0})^+$ distinct A_α's which all have the same intersection with Z, and which therefore all have the same intersection with each other. $\square$

Lemma 20.6. *Let P be a forcing notion, G a generic ideal of P, A a set in $\mathbf{M}$, $B \in \mathbf{M}[G]$ a subset of A, and κ an infinite cardinal in $\mathbf{M}$. Suppose $\mathbf{M} \models$ "A and every join-antichain in P have cardinality at most κ". Then* there is a P-name τ such that $\tau^G = B$ and $\mathbf{M} \models$ "τ is generated by at most κ elements, each of the form $\langle \check{x}, p \rangle$ for some $x \in A$ and $p \in P$".

Proof. Let σ be a P-name that evaluates to B in $\mathbf{M}[G]$. Working in $\mathbf{M}$, for each $x \in A$ let S_x be the set of $p \in P$ which force $\check{x} \in \sigma$ and let S'_x be a maximal join-antichain in S_x. Then define $\tau_0 = \{ \langle \check{x}, p \rangle \mid x \in A$ and $p \in S'_x \}$ and let τ be the P-name generated by τ_0. It is clear that $\mathbf{M} \models \mathrm{card}(\tau_0) \leq \kappa$. We must show that $\tau^G = B$. Certainly, if $\langle \check{x}, p \rangle \in \tau_0$ and $p \in G$ then $x \in \sigma^G = B$. So $\tau^G \subseteq B$. Conversely, if $x \in B$ then some element p of G forces $\check{x} \in \sigma$. Then every extension of p lies in S_x, which implies that every element of G has an extension in S_x, so that G intersects S'_x by Lemma 10.1. This implies that $x \in \tau^G$, so we have $\tau^G = B$. □

We will force $(\forall S) \Diamond_S$ using sets of triples $\{ \langle \beta, \gamma, h^\gamma_\beta \rangle \mid \beta < \alpha_\gamma$ and $\gamma \in X \}$ where X is a countable subset of $(2^{\aleph_0})^+$ and, for each γ, the sequence $\{ h^\gamma_\beta \mid \beta < \alpha_\gamma \}$ is a partial vertex selection from the complete $\aleph_1$-$\aleph_1$ tree as in Theorem 17.2. We call such a set a *partial vertex selection from* $(2^{\aleph_0})^+$ *complete* $\aleph_1$-$\aleph_1$ *trees*.

Theorem 20.7. *Let P be the set of all $p \in \mathbf{M}$ such that $\mathbf{M} \models$ "p is a partial vertex selection from $(2^{\aleph_0})^+$ complete $\aleph_1$-$\aleph_1$ trees" and let G be a generic ideal of P. Then $\mathbf{M}[G] \models$ "$\Diamond_S$ holds for all stationary $S \subseteq \aleph_1$".*

Proof. Fix $S \in \mathbf{M}[G]$ such that $\mathbf{M}[G] \models$ "S is a stationary subset of $\aleph_1$". By an argument similar to the one in the proof of Theorem 13.5, but using Lemma 20.5 instead of Lemma 13.4, we have that $\mathbf{M} \models$ "every join-antichain in P has cardinality at most $2^{\aleph_0}$". Since $S \subseteq \aleph_1^{\mathbf{M}[G]} = \aleph_1^{\mathbf{M}}$ (P is ω-closed), we can use Lemma 20.6 to find a P-name τ for S which is generated by a set τ_0 of pairs of the form $\langle \check{\alpha}, p \rangle$ such that $\mathbf{M} \models \mathrm{card}(\tau_0) \leq 2^{\aleph_0}$.

In $\mathbf{M}$, at most $2^{\aleph_0}$ elements of P appear in τ_0, and each element of P selects from only countably many trees. So there exists an index γ_0 such that τ_0 lives on the set $P_1 \subset P$ of partial vertex selections from trees indexed by $\gamma \neq \gamma_0$. Letting P_2 be the set of partial vertex selections from the single tree indexed by γ_0, we then have $P = P_1 \cdot P_2$, so Theorem 20.4 applies and we can write $G = G_1 \cdot G_2$. Evidently $S \in \mathbf{M}[G_1]$. Since S is stationary in $\mathbf{M}[G]$ and every club set in $\mathbf{M}[G_1]$ is also club in $\mathbf{M}[G]$, S must be stationary in $\mathbf{M}[G_1]$. Thus Lemma 20.2 implies that $\mathbf{M}[G] \models \Diamond_S$. Since S was arbitrary, we have shown that $\mathbf{M}[G] \models (\forall S) \Diamond_S$. □

Metatheorem 20.8. *If ZFC is consistent, then so is ZFC $+ (\forall S) \Diamond_S$.*

Chapter 21

Whitehead's Problem, I*

Let $\mathcal{A}$ be an abelian group. An *extension of $\mathcal{A}$ by $\mathbb{Z}$* is a surjective homomorphism π from an abelian group $\mathcal{B}$ into $\mathcal{A}$ whose kernel is isomorphic to $\mathbb{Z}$, i.e., it is a short exact sequence

$$0 \longrightarrow \mathbb{Z} \longrightarrow \mathcal{B} \longrightarrow \mathcal{A} \longrightarrow 0.$$

It is *trivial* if $\mathcal{B} \cong \mathcal{A} \oplus \mathbb{Z}$ in a way that identifies π with the projection onto the first summand. Equivalently, π is trivial if there is a homomorphism $\rho : \mathcal{A} \to \mathcal{B}$ such that $\pi \circ \rho = \mathrm{id}_{\mathcal{A}}$, since this gives rise to the decomposition $\mathcal{B} = \rho(\mathcal{A}) \oplus \ker \pi \cong \mathcal{A} \oplus \mathbb{Z}$. We say that ρ *splits* π.

An abelian group $\mathcal{A}$ is *free* if it has a *basis*, a set of elements $S \subset \mathcal{A}$ that is linearly independent over $\mathbb{Z}$ and whose $\mathbb{Z}$-linear span equals $\mathcal{A}$. If $\mathcal{A}$ is free then every extension of $\mathcal{A}$ by $\mathbb{Z}$ is trivial: given a basis S, we can construct a splitting of $\pi : \mathcal{B} \to \mathcal{A}$ by choosing $\rho(a)$ arbitrarily in $\pi^{-1}(a)$ for each $a \in S$, and then extending to $\mathcal{A}$ by $\mathbb{Z}$-linearity. *Whitehead's problem* asks whether this property characterizes free abelian groups: if $\mathcal{A}$ has only trivial extensions by $\mathbb{Z}$, is it free?

The answer is yes if $\mathcal{A}$ is finitely generated, as an easy consequence of the structure theorem for finitely generated abelian groups. More is true:

Theorem 21.1. *Let $\mathcal{A}$ be a countable abelian group all of whose extensions by $\mathbb{Z}$ are trivial. Then $\mathcal{A}$ is free.*

This is a theorem of ZFC. We omit the proof since it is essentially an easier version of the proof of Theorem 21.5 below. Instead, we will use this positive answer for countable groups to show that $(\forall S)\diamondsuit_S$ implies the same conclusion when $\mathcal{A}$ has cardinality $\aleph_1$, a result due to Shelah. Shelah actually generalized the argument for $\aleph_1$ to arbitrary uncountable cardinalities, proving that a complete positive answer to Whitehead's problem is relatively consistent with ZFC.

We start with two easy lemmas about abelian groups.

Lemma 21.2. *Let $\mathcal{A}$ be an abelian group, let $\mathcal{A}_0$ be a subgroup of $\mathcal{A}$, and let $\pi_0 : \mathcal{B}_0 \to \mathcal{A}_0$ be an extension of $\mathcal{A}_0$ by $\mathbb{Z}$. Then π_0 can be embedded in an extension $\pi : \mathcal{B} \to \mathcal{A}$ of $\mathcal{A}$ by $\mathbb{Z}$.*

Proof. Using Zorn's lemma, it is enough to consider the case that $\mathcal{A}$ is generated by $\mathcal{A}_0$ and one additional element a. Assume this. If $na \notin \mathcal{A}_0$ for any $n > 0$ then $\mathcal{A} \cong \mathcal{A}_0 \oplus \mathbb{Z}$ and the proof is easy. Otherwise, let n be the least positive integer such that $na \in \mathcal{A}_0$, fix $b \in \pi_0^{-1}(na)$, and let $\mathcal{B} = (\mathcal{B}_0 \oplus \mathbb{Z})/\mathcal{C}$ where $\mathcal{C}$ is the cyclic subgroup of $\mathcal{B}_0 \oplus \mathbb{Z}$ generated by $b \oplus n$. Define $\pi : \mathcal{B} \to \mathcal{A}$ by $\pi((x \oplus m) + \mathcal{C}) = \pi_0(x) - ma$. It is routine to check that π is well-defined and surjective, that embedding $\mathcal{B}_0$ in $\mathcal{B}$ by the map $x \mapsto (x \oplus 0) + \mathcal{C}$ makes π_0 agree with the restriction of π to the embedded copy of $\mathcal{B}_0$, and that

$$\ker \pi = \big\{ (z \oplus 0) + \mathcal{C} \mid z \in \ker \pi_0 \big\},$$

so that $\ker \pi \cong \ker \pi_0 \cong \mathbb{Z}$. $\square$

Since Whitehead's problem has a positive answer for countable groups (Theorem 21.1), it follows from Lemma 21.2 that if $\mathcal{A}$ has only trivial extensions by $\mathbb{Z}$ then all countable subgroups of $\mathcal{A}$ are free. For if $\mathcal{A}_0 \subseteq \mathcal{A}$ is countable and $\pi_0 : \mathcal{B}_0 \to \mathcal{A}_0$ is an extension by $\mathbb{Z}$, then the extension $\pi : \mathcal{B} \to \mathcal{A}$ provided by Lemma 21.2 splits, and this splitting restricts to a splitting of π_0. This works because $\pi^{-1}(\mathcal{A}_0) = \mathcal{B}_0$, which follows from the fact that $\ker \pi = \ker \pi_0$.

Lemma 21.3. *Let $\mathcal{A}$ be an abelian group, $\mathcal{A}_0$ a subgroup of $\mathcal{A}$, and $\rho_0 : \mathcal{A}_0 \to \mathcal{A}_0 \oplus \mathbb{Z}$ the standard splitting of the trivial extension of $\mathcal{A}_0$ by $\mathbb{Z}$. Suppose $\mathcal{A}/\mathcal{A}_0$ has a nontrivial extension by $\mathbb{Z}$. Then the trivial extension of $\mathcal{A}_0$ by $\mathbb{Z}$ can be embedded in an extension of $\mathcal{A}$ by $\mathbb{Z}$ no splitting of which restricts to ρ_0 on $\mathcal{A}_0$.*

Proof. Let $\mathcal{A}' = \mathcal{A}/\mathcal{A}_0$ and let $\pi' : \mathcal{B}' \to \mathcal{A}'$ be a nontrivial extension of $\mathcal{A}'$ by $\mathbb{Z}$. Let $\mathcal{B}$ be the subgroup of $\mathcal{A} \oplus \mathcal{B}'$ consisting of all elements of the form $x \oplus y$ such that $\pi'(y) = x + \mathcal{A}_0$, and define $\pi : \mathcal{B} \to \mathcal{A}$ by $\pi(x \oplus y) = x$. It is routine to check that π is surjective, that $\ker \pi \cong \ker \pi' \cong \mathbb{Z}$, and that π restricted to $\mathcal{A}_0 \oplus \ker \pi$ is a trivial extension of $\mathcal{A}_0$ by $\mathbb{Z}$. If ρ_0 embedded in a splitting $\rho : \mathcal{A} \to \mathcal{B}$ of π, so that we had $\rho(x) = x \oplus 0$ for all $x \in \mathcal{A}_0$, then we could define a splitting ρ' of π' by setting $\rho'(x + \mathcal{A}_0) = y$ where $\rho(x) = x \oplus y$. But this is impossible because π' is nontrivial. $\square$

The next lemma contains the substance of the proof. As usual, $\mathcal{A}_{<\alpha} = \bigcup_{\beta<\alpha} \mathcal{A}_\beta$. Say that an increasing sequence $(\mathcal{A}_\alpha)$ is *strictly increasing* if $\mathcal{A}_{<\alpha} \neq \mathcal{A}_\alpha$ for all α.

Lemma 21.4. *Assume* $(\forall S) \Diamond_S$ *and let* $\mathcal{A}$ *be an abelian group all of whose extensions by* $\mathbb{Z}$ *are trivial. Suppose* $\mathcal{A} = \bigcup_{\alpha<\aleph_1} \mathcal{A}_\alpha$ *for a strictly increasing transfinite sequence of countably infinite subgroups* $\mathcal{A}_\alpha$. *Then* $\mathcal{A}_\alpha/\mathcal{A}_{<\alpha}$ *is free for all* α *in some club subset of* $\aleph_1$.

Proof. We can assume $\mathcal{A}_0$ is not the zero group. Construct a tree $\mathcal{T}$ of height $\aleph_1$ by letting the vertices at level α be all the functions $f : \mathcal{A}_{<\alpha} \to \mathbb{N} \times \alpha$ such that $f(\mathcal{A}_\beta) \subseteq \mathbb{N} \times (\beta + 1)$ for all $\beta < \alpha$. $\mathcal{T}$ is ordered by inclusion. Since $\Diamond$ implies CH, this is a complete $\aleph_1$-$\aleph_1$ tree. Suppose the lemma fails; then the set $S \subseteq \aleph_1$ of values of α for which $\mathcal{A}_\alpha/\mathcal{A}_{<\alpha}$ is not free is stationary. Use $\Diamond_S$ to choose a transfinite sequence of vertices h_α of $\mathcal{T}$ for $\alpha \in S$.

We now recursively construct a transfinite increasing sequence of countable abelian groups $\mathcal{C}_\alpha$, $\alpha < \aleph_1$, such that the underlying set of $\mathcal{C}_\alpha$ is $\mathbb{N} \times (\alpha + 1)$, together with surjective homomorphisms $\pi_\alpha : \mathcal{C}_\alpha \to \mathcal{A}_\alpha$ satisfying $\ker \pi_\alpha \cong \mathbb{Z}$ for all α and $\pi_\alpha|_{\mathcal{C}_\beta} = \pi_\beta$ for all $\beta < \alpha$. Given $\mathcal{C}_{<\alpha}$ and $\pi_{<\alpha} : \mathcal{C}_{<\alpha} \to \mathcal{A}_{<\alpha}$ (defined in the obvious way), if $\alpha \notin S$ or h_α fails to split $\pi_{<\alpha}$ then we use Lemma 21.2 to embed $\pi_{<\alpha}$ in an extension $\pi_\alpha : \mathcal{C}_\alpha \to \mathcal{A}_\alpha$ of $\mathcal{A}_\alpha$ by $\mathbb{Z}$ and we identify the underlying set of $\mathcal{C}_\alpha$ with $\mathbb{N} \times (\alpha+1)$ in any way that extends the previous identification of the underlying set of $\mathcal{C}_{<\alpha}$ with $\mathbb{N} \times \alpha$. Otherwise, if $\alpha \in S$ and h_α splits $\pi_{<\alpha}$, then we use the fact that $\mathcal{A}_\alpha/\mathcal{A}_{<\alpha}$ is not free (since $\alpha \in S$) plus the positive answer to Whitehead's problem for countable groups to find a nontrivial extension of $\mathcal{A}_\alpha/\mathcal{A}_{<\alpha}$ by $\mathbb{Z}$. Then we apply Lemma 21.3 to get an extension $\pi_\alpha : \mathcal{C}_\alpha \to \mathcal{A}_\alpha$ of $\mathcal{A}_\alpha$ by $\mathbb{Z}$ which restricts to $\pi_{<\alpha}$ on $\mathcal{C}_{<\alpha}$ but such that h_α does not embed in a splitting of π_α. Again, we can arrange that the underlying set of $\mathcal{C}_\alpha$ is $\mathbb{N} \times (\alpha + 1)$. This completes the construction.

Let $\mathcal{C} = \bigcup_{\alpha<\aleph_1} \mathcal{C}_\alpha$ and define $\pi : \mathcal{C} \to \mathcal{A}$ by letting $\pi|_{\mathcal{C}_\alpha} = \pi_\alpha$ for all α. This is an extension of $\mathcal{A}$ by $\mathbb{Z}$, and since $\mathcal{A}$ has only trivial extensions by $\mathbb{Z}$ there is a splitting $\rho : \mathcal{A} \to \mathcal{C}$ of π. We have $\rho(\mathcal{A}_\alpha) \subset \mathcal{C}_\alpha$ for all α since $\pi^{-1}(\mathcal{A}_\alpha) = \mathcal{C}_\alpha$. Therefore by $\Diamond_S$ there exists $\alpha \in S$ such that $\rho|_{\mathcal{A}_{<\alpha}} = h_\alpha$. By the construction $\rho|_{\mathcal{A}_{<\alpha}}$ cannot embed in a splitting of π_α, yet it does embed in the splitting $\rho|_{\mathcal{A}_\alpha}$, a contradiction. We conclude that $\mathcal{A}_\alpha/\mathcal{A}_{<\alpha}$ cannot fail to be free on a stationary set. $\quad\square$

Theorem 21.5. *Assume* $(\forall S) \Diamond_S$ *and let* $\mathcal{A}$ *be a group of cardinality* $\aleph_1$ *all* *of whose extensions by $\mathbb{Z}$ are trivial. Then $\mathcal{A}$ is free.*

Proof. First, we claim that for any countable subset S of $\mathcal{A}$ there exists a countably infinite subgroup $\mathcal{B} \subset \mathcal{A}$ that contains S and such that for any countable subgroup $\tilde{\mathcal{B}} \subset \mathcal{A}$ that contains $\mathcal{B}$ the quotient $\tilde{\mathcal{B}}/\mathcal{B}$ is free. Assuming the claim fails for some S, construct a strictly increasing transfinite sequence of countably infinite subgroups $(\tilde{\mathcal{A}}_\alpha)$ by finding $\tilde{\mathcal{A}}_\alpha$ ($= \tilde{\mathcal{B}}$) which falsifies the claim for $\tilde{\mathcal{A}}_{<\alpha}$ ($= \mathcal{B}$). Then by Lemma 21.2 the group $\bigcup_{\alpha < \aleph_1} \tilde{\mathcal{A}}_\alpha \subseteq \mathcal{A}$ satisfies the hypotheses of Lemma 21.4, which is absurd because $\tilde{\mathcal{A}}_\alpha/\tilde{\mathcal{A}}_{<\alpha}$ is not free for any $\alpha > 0$. This proves the claim. (Ordinary diamond would suffice for this part of the argument, since we only need a single quotient to be free, not a club sequence of quotients.)

Now enumerate the elements of $\mathcal{A}$ as $\{x_\alpha \mid \alpha < \aleph_1\}$ and use the claim to find a strictly increasing transfinite sequence of countably infinite subgroups $\mathcal{A}_\alpha$ of $\mathcal{A}$ such that, for each α, $\mathcal{A}_\alpha$ contains $\mathcal{A}_{<\alpha} \cup \{x_\alpha\}$ and the quotient $\tilde{\mathcal{B}}/\mathcal{A}_\alpha$ is free for any countable subgroup $\tilde{\mathcal{B}}$ containing $\mathcal{A}_\alpha$. By Lemma 21.4 there is a club sequence (α_β) with the property that $\mathcal{A}_{\alpha_\beta}/\mathcal{A}_{<\alpha_\beta}$ is free for all β. Define $\mathcal{A}'_\beta = \mathcal{A}_{\alpha_\beta}$. Then for all β the quotient $\mathcal{A}'_{\beta+1}/\mathcal{A}'_\beta$ is free since every $\tilde{\mathcal{B}}/\mathcal{A}_\alpha$ is free, and if β is a limit ordinal then since the sequence (α_β) is closed we have $\mathcal{A}'_{<\beta} = \mathcal{A}_{<\alpha_\beta}$, so that $\mathcal{A}'_\beta/\mathcal{A}'_{<\beta} = \mathcal{A}_{\alpha_\beta}/\mathcal{A}_{<\alpha_\beta}$ is also free. But whenever a quotient is free the quotient map splits (for the same reason any extension of a free group by $\mathbb{Z}$ splits), so for each β we can find a countable subgroup $\mathcal{B}_\beta$ of $\mathcal{A}'_\beta$ such that $\mathcal{A}'_\beta = \mathcal{B}_\beta \oplus \mathcal{A}'_{<\beta}$. Thus $\mathcal{A} = \bigoplus_{\beta < \aleph_1} \mathcal{B}_\beta$. As we know that every countable subgroup of $\mathcal{A}$ is free (see the comment following Lemma 21.2), we have expressed $\mathcal{A}$ as a direct sum of free abelian groups, and therefore $\mathcal{A}$ is free. $\square$

Chapter 22

Iterated Forcing

Working in ZFC$^+$, we have developed a procedure for enlarging a countable transitive model $\mathbf{M}$ of ZFC to another countable transitive model $\mathbf{M}[G]$ using a forcing notion for $\mathbf{M}$. But now the same procedure can be applied to $\mathbf{M}[G]$ using a forcing notion for $\mathbf{M}[G]$. We already saw an example of this in Chapter 20, although in that case both forcing notions belonged to $\mathbf{M}$. Iterating forcing constructions can be a useful technique, but as we will show, it does not actually create any greater generality.

We start by considering a two-stage process which involves a forcing notion P and a P-name π for a forcing notion, with the idea of first forcing with P, then evaluating π and forcing with it. It is convenient to assume that P is rooted, i.e., $\emptyset \in P$. The corresponding condition on π is $\langle \emptyset, \emptyset \rangle \in \pi$; we say that π is a *rooted P-name*.

Definition 22.1. Let P be a rooted forcing notion for $\mathbf{M}$ and let π be a rooted P-name. For each $\langle \tau, p \rangle \in \pi$ define $p * \tau \subseteq \pi$ by

$$p * \tau = \big\{ \langle \sigma, q \rangle \in \pi \mid q \subseteq p \text{ and } p \Vdash \sigma \subseteq \tau \big\},$$

and let $P * \pi$ be the set

$$P * \pi = \big\{ p * \tau \mid \langle \tau, p \rangle \in \pi \big\}.$$

Lemma 22.2. *Let P be a rooted forcing notion for $\mathbf{M}$, let π be a rooted P-name, and let $\langle \tau, p \rangle, \langle \tau', p' \rangle \in \pi$. Then $p * \tau \subseteq p' * \tau'$ if and only if $p \subseteq p'$ and $p' \Vdash \tau \subseteq \tau'$.*

Proof. Observe that $\langle \tau, p \rangle \in p * \tau$. So $p * \tau \subseteq p' * \tau'$ implies that $\langle \tau, p \rangle \in p' * \tau'$, which immediately yields the forward direction of the lemma. Conversely, if $p \subseteq p'$ and $p' \Vdash \tau \subseteq \tau'$ then any $\langle \sigma, q \rangle \in \pi$ satisfying $q \subseteq p$ and $p \Vdash \sigma \subseteq \tau$ will also satisfy $q \subseteq p'$ and $p' \Vdash \sigma \subseteq \tau \subseteq \tau'$, so that $p * \tau \subseteq p' * \tau'$. $\square$

Theorem 22.3. *Let P be a rooted forcing notion for $\mathbf{M}$, let π be a rooted P-name, and let G be an ideal of $P * \pi$ which is generic relative to $\mathbf{M}$. Then*

$$G_1 = \big\{ p \mid p * \tau \in G \text{ for some } \tau \big\}$$

is an ideal of P which is generic relative to $\mathbf{M}$ and

$$G_2 = \big\{ \tau^{G_1} \mid p * \tau \in G \text{ for some } p \big\}$$

is an ideal of π^{G_1} which is generic relative to $\mathbf{M}[G_1]$. We have $\mathbf{M}[G] = \mathbf{M}[G_1][G_2]$.

Proof. Observe that $p * \tau \in G$ for some τ if and only if $p * \emptyset \in G$. It straightforwardly follows that G_1 is an ideal. To show it is generic, let $D_1 \subseteq P$ be a dense set that lies in $\mathbf{M}$. Then

$$D = \big\{ p * \tau \mid \langle \tau, p \rangle \in \pi \text{ and } p \in D_1 \big\}$$

is a dense subset of $P * \pi$ that also lies in $\mathbf{M}$. Thus there exists $p*\tau \in G \cap D$, so that $p \in G_1 \cap D_1$. So G_1 is generic relative to $\mathbf{M}$.

We verify that G_2 is an ideal of π^{G_1}. First note that if $\tau^{G_1} \in G_2$ with $p * \tau \in G$, then $p \in G_1$ and so $\tau^{G_1} \in \pi^{G_1}$. Thus $G_2 \subseteq \pi^{G_1}$. Now any element of π^{G_1} that is contained in τ^{G_1} must equal σ^{G_1} for some $\langle \sigma, q \rangle \in \pi$ with $q \in G_1$. By directedness of G_1 we can assume $q \supseteq p$ and $q \Vdash \sigma \subseteq \tau$. Then since $q*\emptyset$ and $p*\tau$ both lie in G, there must exist $p' * \tau' \in G$ such that $q \subseteq p'$ and $p * \tau \subseteq p' * \tau'$, and Lemma 22.2 then implies that $q * \sigma \subseteq p' * \tau'$. So $q * \sigma \in G$, and hence $\sigma^{G_1} \in G_2$. We conclude that G_2 is downward stable. Next, any $p_1 * \tau_1$ and $p_2 * \tau_2$ in G have a common extension $p * \tau$ in G, and then $p \Vdash \tau_1 \cup \tau_2 \subseteq \tau$, so that $\tau_1^{G_1} \cup \tau_2^{G_1} \subseteq \tau^{G_1} \in G_2$. So G_2 is also directed.

To see that G_2 is generic relative to $\mathbf{M}[G_1]$, suppose σ^{G_1} is a dense subset of π^{G_1}. Define

$$D = \big\{ p * \tau \mid \langle \tau, p \rangle \in \pi \text{ and } p \Vdash \tau \in \sigma \big\}.$$

Then $D \in \mathbf{M}$, and we claim that every element of G lies below an element of D. For if $p * \tau \in G$ then $\tau^{G_1} \in \pi^{G_1}$, so there exists $\langle \tau', p' \rangle \in \pi$ such that $p' \in G_1$ and $\tau^{G_1} \subseteq \tau'^{G_1} \in \sigma^{G_1}$. By directedness of G_1 we can assume $p' \supseteq p$ and $p' \Vdash \tau \subseteq \tau' \in \sigma$, and then $p' * \tau'$ belongs to D and contains $p*\tau$. This proves the claim.

By Lemma 10.1, there must exist $p*\tau \in G \cap D$. But then $\tau^{G_1} \in G_2 \cap \sigma^{G_1}$, so G_2 meets σ^{G_1}. We have shown that G_2 is generic relative to $\mathbf{M}[G_1]$.

Finally, it is clear that $G_1, G_2 \in \mathbf{M}[G]$; working in $\mathbf{M}[G]$, a straightforward double induction on name rank shows that $\mathbf{M}[G_1][G_2] \subseteq \mathbf{M}[G]$.

Conversely, let $G_1 * G_2$ be the set of all $p * \tau \in P * \pi$ such that $p \in G_1$ and $\tau^{G_1} \in G_2$. We claim that $G = G_1 * G_2$; this will show that G belongs to $\mathbf{M}[G_1][G_2]$ and hence that $\mathbf{M}[G] \subseteq \mathbf{M}[G_1][G_2]$. In one direction, it is immediate that $p * \tau \in G$ implies $p * \tau \in G_1 * G_2$. Conversely, given any $p * \tau \in P * \pi$ such that $p \in G_1$ and $\tau^{G_1} \in G_2$, we have $\tau^{G_1} = \sigma^{G_1}$ for some $q * \sigma \in G$, and by directedness of G_1 we can assume $q \supseteq p$ and $q \Vdash \tau = \sigma$. But then $p * \tau \subseteq q * \sigma$, so we must have $p * \tau \in G$. This proves the reverse inclusion, and we conclude that $G_1 * G_2 = G$. $\qquad\square$

The definition of transfinitely iterated forcing is a little technical. We can simplify it by modifying the definition of $P * \pi$ so that it contains P. First, we "disjointify" P and π by setting $\tilde{\tau}_0 = \tau_0 \times \{P\}$ for any subset τ_0 of π; this ensures that $p \cap \tilde{\tau}_0 = \emptyset$ for all $p \in P$. Then for $\langle \tau, p \rangle \in \pi$ define $p \tilde{*} \tau = p \cup (\widetilde{p * \tau})$, observe that $p \tilde{*} \tau \subseteq p' \tilde{*} \tau'$ if and only if $p * \tau \subseteq p' * \tau'$, and set

$$ P \tilde{*} \pi = P \cup \left\{ p \tilde{*} \tau \mid \langle \tau, p \rangle \in \pi \right\}. $$

For any $p \in P$ we have $p \subset p \tilde{*} \emptyset$, so that P lies below the rest of $P \tilde{*} \pi$ and thus forcing with $P \tilde{*} \pi$ is effectively equivalent to forcing with $P * \pi$.

Definition 22.4. Let α be an ordinal in $\mathbf{M}$. An α-*stage finite support iterated forcing construction* consists of two sequences, $\{P_\beta \mid 0 \le \beta < \alpha\}$ and $\{\pi_\beta \mid 1 \le \beta < \alpha\}$, such that $P_0 = \{\emptyset\}$ and

(i) π_β is a rooted $P_{<\beta}$-name
(ii) $P_\beta = P_{<\beta} \tilde{*} \pi_\beta$

for all $1 \le \beta < \alpha$, where $P_{<\beta} = \bigcup_{\gamma < \beta} P_\gamma$. We also define $P^* = P_{<\alpha}$.

Definition 22.4 is a "finite support" construction because each element of P^* ends up being a disjoint union $\bigcup_{1 \le \beta < \alpha} \tilde{\tau}_\beta$ with each $\tau_\beta \subseteq \pi_\beta$ and such that $\tau_\beta = \emptyset$ for all but finitely many β. This holds at limit stages because each element of $P_{<\beta}$ belongs to P_γ for some $\gamma < \beta$.

Theorem 22.5. *Given an α-stage finite support iterated forcing construction as in Definition 22.4, let G be an ideal of P^* which is generic relative to $\mathbf{M}$. Then for each $\beta < \alpha$ the set $G_\beta = G \cap P_{<\beta}$ is an ideal of $P_{<\beta}$ which is generic relative to $\mathbf{M}$, and the set $G'_\beta = \{\tau^{G_\beta} \mid p \tilde{*} \tau \in G_{\beta+1}$ for some $p \in P_{<\beta}\}$ is an ideal of $\pi_\beta^{G_\beta}$ which belongs to $\mathbf{M}[G_{\beta+1}]$ and is generic relative to $\mathbf{M}[G_\beta]$.*

The proof is similar to the proof of Theorem 22.3, so we omit it. We show that iterated forcing preserves the c.c.c. property.

Lemma 22.6. *Let P be a rooted forcing notion in $\mathbf{M}$ and let π be a rooted P-name. Suppose $\mathbf{M} \models$ "P is c.c.c." and $\emptyset \Vdash$ "π is c.c.c." Then $\mathbf{M} \models$ "$P * \pi$ is c.c.c."*

Proof. Let γ be an ordinal in $\mathbf{M}$ and suppose $\{p_\alpha * \tau_\alpha \mid \alpha < \gamma\} \in \mathbf{M}$ is a family of pairwise incompatible elements of $P * \pi$. Let σ be the P-name generated by the pairs $\langle \check{\alpha}, p_\alpha \rangle$ for $\alpha < \gamma$. For any generic ideal G of P, we claim that the elements τ_α^G of π^G for $\alpha \in \sigma^G$, (i.e., for $p_\alpha \in G$) are pairwise incompatible. Indeed, if τ_α^G and τ_β^G had a common extension τ^G for some $\alpha, \beta \in \sigma^G$, where $\langle \tau, p \rangle \in \pi$ with $p \in G$, then by directedness we could assume $p \supseteq p_\alpha, p_\beta$ and $p \Vdash \tau_\alpha \cup \tau_\beta \subseteq \tau$, but this would imply that $p_\alpha * \tau_\alpha, p_\beta * \tau_\beta \subseteq p * \tau$, a contradiction. This proves the claim.

Since $\emptyset$ forces "π is c.c.c." it must therefore force "σ is countable". Now, working in $\mathbf{M}$, for each $\alpha < \gamma$ choose $q_\alpha \in P$, if one exists, which forces $\check{\alpha} = \sup \sigma$. Then these q_α are pairwise incompatible, so there are only countably many of them. It follows that $\mathbf{M} \models$ "β is countable" where β is the least ordinal greater than all the α which are forced by some element of P to equal $\sup \sigma$. But $p_\alpha \Vdash \check{\alpha} \leq \sigma$, so some extension of p_α forces $\check{\alpha}' = \sup \sigma$ for some $\alpha' \geq \alpha$. Thus $\gamma \leq \beta$, so $\mathbf{M} \models$ "γ is countable". $\qquad\square$

Lemma 22.7. *Given an α-stage finite support iterated forcing construction as in Definition 22.4, suppose $\emptyset$ forces "π_β is c.c.c." (relative to $P_{<\beta}$) for all $\beta < \alpha$. Then $\mathbf{M} \models$ "P^* is c.c.c."*

Proof. Working in $\mathbf{M}$, we prove that $P_{<\beta}$ is c.c.c. by induction on β. At successor stages this follows from Lemma 22.6. Now suppose β is a limit ordinal and let X be an uncountable subset of $P_{<\beta}$. Recall that each $p \in P_{<\beta}$ is a disjoint union $\bigcup_{1 \leq \gamma < \beta} \tilde{\tau}_\gamma$ with each $\tau_\gamma \subseteq \pi_\gamma$ and $\tau_\gamma = \emptyset$ for all but finitely many values of γ. Let the *support* of p be $\{\gamma < \beta \mid \tau_\gamma \neq \emptyset\}$. Then by the Δ-system lemma (Lemma 13.4) we can find an uncountable subset $Y \subseteq X$ and a set R such that the intersection of the supports of any two distinct elements of Y equals R. But then $R \subseteq \gamma$ for some $\gamma < \beta$ (since β is a limit ordinal), and we know P_γ is c.c.c.; therefore the intersections of some pair of distinct elements of Y with P_γ must be compatible. But this implies that these two elements of Y are themselves compatible; this is proven by induction, based on the fact that if $\langle \tau, p \rangle \in \pi_\gamma$ and $\langle \tau', p \rangle \in \pi_{\gamma'}$ where $\gamma < \gamma'$, then $p \,\tilde{*}\, \tau, p \,\tilde{*}\, \tau' \subseteq (p \,\tilde{*}\, \tau) \,\tilde{*}\, \tau'$. Thus $P_{<\beta}$ is c.c.c. $\qquad\square$

Chapter 23

Martin's Axiom

Definitions 8.1 and 8.2 can be generalized to arbitrary partially ordered sets. Let $\mathcal{P}$ be a poset. An *extension* of $p \in \mathcal{P}$ is any $q \geq p$; p and q are *compatible* if they have a common extension; a subset D of $\mathcal{P}$ is *dense* if every $p \in \mathcal{P}$ has an extension in D; and an *ideal* of $\mathcal{P}$ is a subset $G \subseteq \mathcal{P}$ satisfying (1) if $q \in G$ and $q \geq p$ then $p \in G$, and (2) if $p_1, p_2 \in G$ then there exists $q \in G$ with $q \geq p_1, p_2$. Also, we say that $\mathcal{P}$ is c.c.c. if every set of pairwise incompatible elements is countable.

Martin's axiom (MA) is the assertion that if $\mathcal{P}$ is a nonempty c.c.c. poset and $\{D_\alpha\}$ is a family of fewer than $2^{\aleph_0}$ dense subsets of $\mathcal{P}$, then there is an ideal of $\mathcal{P}$ that intersects every D_α. Although it uses the language of forcing, this is a statement of ordinary mathematics.

In general it is too much to ask for an ideal that meets $2^{\aleph_0}$ dense sets; we can prove in ZFC that there are counterexamples to this stronger statement. For instance, let $\mathcal{P}$ be an infinite binary tree and for each branch ϕ of $\mathcal{P}$ let D_ϕ be the set of vertices not belonging to ϕ. Then each D_ϕ is dense, but any ideal must sit inside a single branch, so there is some D_ϕ which it does not meet. In this example $\mathcal{P}$ is c.c.c. simply because it is countable.

At the other extreme, by adapting the proof of Theorem 8.3 we can also prove in ZFC that for any countable family of dense sets in any nonempty poset there is an ideal that intersects them all. We do not even need to assume that the poset is c.c.c. This shows that CH implies MA. But, as we will see, MA is also relatively consistent with $\neg$ CH. Specifically, we will force MA + "$2^{\aleph_0} = \aleph_2$". The idea of the proof is to execute an $\aleph_2$-stage iterated forcing construction where at each stage we add an ideal that intersects all currently available dense subsets of some c.c.c. poset. In the ground model where the construction begins there will be $\aleph_2$ posets we want to handle, and at each stage of the construction we potentially introduce $\aleph_2$ new posets to handle, but this means that through $\aleph_2$ stages there will only be $\aleph_2^2 = \aleph_2$ posets to handle in total, and by arranging them carefully we are able to do this in $\aleph_2$ steps. In fact, we have enough time to force with each poset an unbounded number of times, so if we can show that any family of $\aleph_1$ dense sets that appears in the final model must already have appeared at one of the approximating stages, then an ideal would subsequently have been added that meets them all.

We start by showing that it is sufficient to check MA for a limited family of posets. This result is provable in ZFC.

Lemma 23.1. *Assume* $2^{\aleph_0} \leq \aleph_2$ *and suppose Martin's axiom holds for any c.c.c. partial ordering of* $\aleph_1$. *Then Martin's axiom is true.*

Proof. If $2^{\aleph_0} = \aleph_1$ then MA is automatic. So assume $2^{\aleph_0} = \aleph_2$.

The hypothesis trivially implies that MA holds for any c.c.c. poset whose underlying set has cardinality $\aleph_1$. It therefore also holds for any countable poset, for if $\mathcal{P}$ is countable we can place a copy of $\aleph_1$ with its usual ordering below it and then invoke the previous fact.

Now let $\mathcal{P}$ be any c.c.c. poset of cardinality at least $\aleph_2$ and let $\{D_\alpha\}$ be a family of at most $\aleph_1$ dense subsets of $\mathcal{P}$. Construct an increasing sequence of subsets X_n of $\mathcal{P}$ as follows. Start by letting X_0 consist of a single element of $\mathcal{P}$, chosen arbitrarily. Given X_n, construct X_{n+1} by adding, for each $p \in X_n$ and each α, an extension of p that belongs to D_α, and for any pair of compatible elements of X_n, a common extension of them. Then let $\mathcal{P}_0 = \bigcup_{n=0}^\infty X_n$, and give it the order inherited from $\mathcal{P}$.

We inductively have $\mathrm{card}(X_n) \leq \aleph_1$ for all n. Thus $\mathrm{card}(\mathcal{P}_0)$ is at most $\aleph_1$, so $\mathcal{P}_0$ has an ideal G_0 which meets each D_α. Then $G = \{p \in \mathcal{P} \mid p \text{ has an extension in } G_0\}$ is an ideal of $\mathcal{P}$ that meets each D_α. □

We also need to be able to convert partial orderings of $\aleph_1$ into forcing notions.

Lemma 23.2. *Let* P *be a rooted c.c.c. forcing notion such that* $\mathbf{M} \models \mathrm{card}(P) \leq \aleph_1$, *let* $p \in P$, *and let* τ *be a* P-*name every element of which has the form* $\langle \mathrm{op}(\check{\alpha}, \check{\beta}), q \rangle$ *for some* $\alpha, \beta \in \aleph_1^{\mathbf{M}}$, *and such that* $p \Vdash$ *"τ is a c.c.c. partial ordering of* $\check{\aleph}_1^{\mathbf{M}}$ *". Then there is a* P-*name* σ *such that*

(i) $\mathbf{M} \models \mathrm{card}(\sigma) = \aleph_1$

(ii) $\emptyset \Vdash$ *"σ is a rooted c.c.c. forcing notion"*

(iii) p *forces that the forcing notion* $\sigma \setminus \{\check{\emptyset}\}$ *ordered by inclusion is isomorphic to the set* $\check{\aleph}_1^{\mathbf{M}}$ *ordered by* τ.

Moreover, the conversion of $\langle \tau, p \rangle$ into σ can be carried out in **M**.

Proof. For each $\beta \in \aleph_1^{\mathbf{M}}$ define

$$\tau_\beta = \left\{ \langle \check{\alpha}, q \rangle \mid \langle \mathrm{op}(\check{\alpha}, \check{\beta}), q \rangle \in \tau \text{ and } q \supseteq p \right\}$$

and then let

$$\sigma = \left\{ \langle \tau_\beta, q \rangle \mid \beta \in \aleph_1^{\mathbf{M}} \text{ and } q \supseteq p \right\} \cup \left\{ \langle \emptyset, q \rangle \mid q \in P \right\}.$$

It is clear that $\mathbf{M} \models \mathrm{card}(\sigma) = \aleph_1$. If G is any generic ideal containing p and $\preceq$ is the ordering of $\aleph_1^{\mathbf{M}}$ to which τ evaluates under G, then we have

$$\sigma^G \setminus \{\emptyset\} = \left\{ \beta^{\preceq} \mid \beta \in \aleph_1^{\mathbf{M}} \right\},$$

where $\beta^{\preceq} = \left\{ \alpha \in \aleph_1^{\mathbf{M}} \mid \alpha \preceq \beta \right\}$. Then $\beta^{\preceq} \subseteq \gamma^{\preceq}$ if and only if $\beta \preceq \gamma$, so $\sigma^G \setminus \{\emptyset\}$ is isomorphic to $\langle \aleph_1^{\mathbf{M}}, \preceq \rangle$. Finally, if G is an arbitrary generic ideal of P then either $p \in G$ and σ^G is c.c.c. by what we just said, or $p \notin G$ and $\sigma^G = \{\emptyset\}$; in either case σ^G is a rooted c.c.c. forcing notion. By inspection, the construction of σ can be executed in **M**. $\qquad\square$

Theorem 23.3. *There is a forcing notion P such that $\mathbf{M}[G] \models MA + $ "$2^{\aleph_0} = \aleph_2$" for any generic ideal G of P.*

Proof. Forcing with the set of bijections between subsets of $\mathcal{P}(\aleph_1)$ and $\aleph_2$ of cardinality at most $\aleph_1$ yields a model which satisfies $2^{\aleph_1} = \aleph_2$; this is proven analogously to Theorem 12.3. For the sake of notational simplicity we will assume $2^{\aleph_1} = \aleph_2$ already holds in **M**.

Work in **M**. We require a bookkeeping function $f : \aleph_2 \to \aleph_2^2$ which is bijective and such that the first coordinate of $f(\beta)$ is at most β, for all $\beta < \aleph_2$. We will recursively define an $\aleph_2$-stage finite support iterated forcing construction. We only need to specify the sequence (π_β), since it determines the sequence (P_β).

Suppose we have determined π_α for all $\alpha < \beta$. We may inductively assume that $P_{<\alpha}$ is c.c.c. and has cardinality at most $\aleph_1$ for every $\alpha \leq \beta$. The c.c.c. property is going to be maintained throughout the construction by Lemmas 22.6 and 22.7, and the cardinality bound will be maintained by ensuring that $\mathrm{card}(\pi_\alpha) \leq \aleph_1$ for all α.

Enumerate the pairs $\langle \tau, p \rangle$ such that τ is a $P_{<\beta}$-name all of whose elements have the form $\langle \mathrm{op}(\check{\alpha}, \check{\beta}), q \rangle$ for some $\alpha, \beta \in \aleph_1^{\mathbf{M}}$ and such that $p \in P_{<\beta}$ forces "τ is a c.c.c. partial ordering of $\aleph_1^{\mathbf{M}}$". Observe that by Lemma 20.6, every c.c.c. partial ordering of $\aleph_1^{\mathbf{M}}$ that appears in $\mathbf{M}[G_\beta]$, for some generic ideal G_β of $P_{<\beta}$, is the value of some such τ. There are $\aleph_2$ possible values of $\langle \tau, p \rangle$, so let $\left\{ \sigma_{\beta,\gamma} \mid \gamma < \aleph_2 \right\}$ enumerate the corresponding σ's from

Lemma 23.2. Then let $\pi_\beta = \sigma_{f(\beta)}$; since the first coordinate of $f(\beta)$ is at most β, we have already defined $\sigma_{f(\beta)}$ at this or some previous stage. This completes the construction of π_β.

Now work in ZFC^+. Let $P = P^*$. Then $\mathbf{M} \models$ "P is c.c.c." by Lemma 22.7. Let G be a generic ideal of P and for each β let $G_\beta = G \cap P_{<\beta}$. Fix a subset $\preceq$ of $(\aleph_1^{\mathbf{M}})^2$ such that $\mathbf{M}[G] \models$ "$\preceq$ is a c.c.c. partial ordering of $\aleph_1^{\mathbf{M}}$"; we claim that $\preceq$ already appears in some $\mathbf{M}[G_\beta]$. To see this, let τ be a P-name for $\preceq$ and, working in $\mathbf{M}[G]$, for each $\langle \alpha', \beta' \rangle \in \aleph_1^{\mathbf{M}}$ such that $\alpha' \preceq \beta'$ choose some $p_{\alpha', \beta'} \in G$ which forces $\mathrm{op}(\check{\alpha}', \check{\beta}') \in \tau$. Since $G = \bigcup_{\beta < \aleph_2^{\mathbf{M}}} G_\beta$ and there are, according to $\mathbf{M}[G]$, only $\aleph_1$ pairs $\langle \alpha', \beta' \rangle$, there must exist $\beta < \aleph_2$ such that each $p_{\alpha', \beta'}$ belongs to G_β. But then $\preceq$ equals

$$\big\{ \langle \alpha', \beta' \rangle \in (\aleph_1^{\mathbf{M}})^2 \mid \text{some } p \in G_\beta \subseteq P \text{ forces } \mathrm{op}(\check{\alpha}', \check{\beta}') \in \tau \big\},$$

and this set can be formed in $\mathbf{M}[G_\beta]$. This proves the claim.

Fix a family $\{D_\alpha\} \in \mathbf{M}[G]$ of at most $\aleph_1^{\mathbf{M}}$ subsets of $\aleph_1^{\mathbf{M}}$ which are dense for the ordering $\preceq$. By running the argument of the last paragraph again, we can assume that this family appears in $\mathbf{M}[G_\beta]$.

We now know that $\langle \aleph_1^{\mathbf{M}}, \preceq \rangle$ is isomorphic to $\sigma_{\beta, \gamma}^G$ for some $\beta, \gamma < \aleph_2^{\mathbf{M}}$ and hence to $\pi_{\beta'}^G = \pi_{\beta'}^{G_{\beta'}}$ where $\beta' = f^{-1}(\beta, \gamma)$. Thus $G_{\beta'}$ adds an ideal of $\langle \aleph_1^{\mathbf{M}}, \preceq \rangle$ which meets every D_α. We conclude that $\mathbf{M}[G] \models$ "Martin's axiom holds for any c.c.c. partial ordering of $\aleph_1$".

Finally, by Lemma 20.6, every subset of $\mathbb{N}$ that appears in $\mathbf{M}[G]$ is the value of a P-name that is countably generated according to $\mathbf{M}$, and there are, according to $\mathbf{M}$, only $\aleph_2^{\aleph_0} = \aleph_2$ such P-names, so $2^{\aleph_0} \leq \aleph_2$ must hold in $\mathbf{M}[G]$. By Lemma 23.1 we conclude that MA holds in $\mathbf{M}[G]$. It then follows that $2^{\aleph_0} = \aleph_1$ cannot be true in $\mathbf{M}[G]$ because we can prove in ZFC that the infinite binary tree has $2^{\aleph_0}$ dense subsets such that no ideal meets all of them, whereas it follows from our construction that any family of $\aleph_1$ dense subsets is met by an ideal. So $\mathbf{M}[G] \models 2^{\aleph_0} = \aleph_2$. $\qquad \square$

Metatheorem 23.4. *If ZFC is consistent, then so is ZFC + MA + "$2^{\aleph_0} = \aleph_2$".*

Chapter 24

Suslin's Problem, II*

Since CH implies MA, one type of application of MA is to weaken uses of CH. That is, we might first prove a result assuming CH and then discover that it can actually be proven from MA. For instance, MA suffices to prove the existence of non-diagonalizable pure states on $\mathcal{B}(H)$ (Theorem 16.4). When this happens it shows that the result in question is relatively consistent with $\neg$ CH. But a more interesting kind of use of MA occurs when CH or $\diamondsuit$ settles a question in one direction and MA $+\ \neg$ CH settles it in the other. We will focus on examples of this phenomenon.

In Chapter 18 we proved that the diamond principle implies Suslin lines exist. The main result of this chapter states that MA $+$ "$2^{\aleph_0} = \aleph_2$" implies Suslin lines do not exist. But first we present some simple consequences of MA relating to measure theory and Baire category.

Theorem 24.1. *Assume MA $+$ "$2^{\aleph_0} = \aleph_2$". Then any union of $\aleph_1$ null subsets of $\mathbb{R}$ is null.*

Proof. Denote Lebesgue measure on $\mathbb{R}$ by m. Fix $\epsilon > 0$ and let $\mathcal{P}$ be the family of all open subsets U of $\mathbb{R}$ such that $m(U) < \epsilon$, ordered by inclusion. We claim that $\mathcal{P}$ is c.c.c.

To see this, define a metric on $\mathcal{P}$ by setting $d(U, V) = m(U \triangle V)$ (the measure of their symmetric difference). This metric is separable because those $U \in \mathcal{P}$ which are finite unions of rational intervals constitute a countable dense subset. Now for any uncountable subset S of $\mathcal{P}$, there exists n such that $S_n = \{U \in S \mid m(U) < \epsilon - \frac{1}{n}\}$ is uncountable. By separability we must then have $d(U, V) < \frac{1}{n}$ for some distinct $U, V \in S_n$, and it follows that $U \cup V \in \mathcal{P}$. We conclude that $\mathcal{P}$ is c.c.c.

Now let $\{N_\alpha\}$ be a family of $\aleph_1$ null subsets of $\mathbb{R}$. For each α let $D_\alpha = \{U \in \mathcal{P} \mid N_\alpha \subseteq U\}$. Each D_α is dense, so by MA there exists an

93

ideal G of $\mathcal{P}$ that meets each D_α. Then the union of the sets in G is an open subset V of $\mathbb{R}$ that contains each N_α. We conclude by showing that $m(V) \leq \epsilon$. To do this, let $r = \sup_{U \in G} m(U) \leq \epsilon$ and choose a sequence (U_n) in G such that $m(U_n) \to r$. By replacing U_n with $U_1 \cup \cdots \cup U_n$ we can assume the sequence (U_n) is increasing. Then $\bigcup U_n$ is an open subset of V whose measure is $r \leq \epsilon$. Now if $m(V) > \epsilon$ then $m(V \setminus \bigcup U_n) > 0$ and there would exist a compact set $C \subseteq V \setminus \bigcup U_n$ of positive measure. But then C would have to be covered by finitely many sets in G, and hence by a single set $U \in G$, and this would imply $m(U_n \cup U) \to m(\bigcup U_n \cup U) \geq r + m(C)$, a contradiction. We conclude that $m(V) \leq \epsilon$. Since $\bigcup N_\alpha \subseteq V$ and $\epsilon > 0$ was arbitrary, this shows that $\bigcup N_\alpha$ is null. $\qquad\square$

Recall that a subset of $\mathbb{R}$ is *meager* if it is a countable union of nowhere-dense sets.

Theorem 24.2. *Assume MA $+$ "$2^{\aleph_0} = \aleph_2$". Then any union of $\aleph_1$ meager subsets of $\mathbb{R}$ is meager.*

Proof. Let $\{C_\alpha \mid \alpha < \aleph_1\}$ be a family of $\aleph_1$ nowhere-dense subsets of $\mathbb{R}$. It will suffice to show that $\bigcup C_\alpha$ is meager.

Define $\mathcal{P}$ to be the family of all finite sequences of ordered pairs $\langle U_1, E_1 \rangle, \ldots, \langle U_n, E_n \rangle$ such that for each i

(i) U_i is a union of finitely many open intervals with rational endpoints
(ii) E_i is a finite subset of $\aleph_1$
(iii) U_i is disjoint from $\bigcup_{\alpha \in E_i} C_\alpha$.

Order $\mathcal{P}$ by setting $p \leq q$ if the sequence q is at least as long as the sequence p and for every $\langle U_i, E_i \rangle$ in p and corresponding $\langle U_i', E_i' \rangle$ in q we have $U_i \subseteq U_i'$ and $E_i \subseteq E_i'$.

Since there are only countably many choices for each U_i, there are only countably many possible finite sequences $U_1, \ldots, U_n$. So any uncountable subset of $\mathcal{P}$ must contain two distinct sequences of the same length which have the same sequence of U_i's. This implies that they are compatible. Thus $\mathcal{P}$ is c.c.c.

For each $\alpha < \aleph_1$, the set D_α of $p \in \mathcal{P}$ which satisfy $\alpha \in E_i$ for some i is dense. Also, for each open interval I with rational endpoints and each $i \in \mathbb{N}$ the set $D_{I,i}$ of $p \in \mathcal{P}$ whose length is at least i and satisfies $U_i \cap I \neq \emptyset$ is dense. So by MA there is an ideal G of $\mathcal{P}$ that intersects every D_α and every $D_{I,i}$.

Now for each $i \in \mathbb{N}$ let V_i be the union of the ith open sets U_i coming from all $p \in G$ of length at least i. Since G meets each $D_{I,i}$ it follows that each V_i is a dense open subset of $\mathbb{R}$. And since G meets each D_α, it follows that for each $\alpha < \aleph_1$ we have $V_i \cap C_\alpha = \emptyset$ for some i. Thus $\bigcap_{i \in \mathbb{N}} V_i$ is a countable intersection of dense open sets whose complement contains $\bigcup_{\alpha < \aleph_1} C_\alpha$. This shows that $\bigcup C_\alpha$ is meager. $\qquad \square$

We return to Suslin's problem. See Chapter 18 for the definition of a Suslin line.

Definition 24.3. A *Suslin tree* is a tree with the following properties:

(i) its height is $\aleph_1$
(ii) every branch is countable
(iii) every antichain is countable.

Lemma 24.4. *If Suslin lines exist then Suslin trees exist.*

Proof. Given a Suslin line, we create a Suslin tree as follows. Start by letting I_0 be the entire line. For $1 \leq \alpha < \aleph_1$, recursively choose I_α to be a bounded closed interval that is nontrivial, i.e., is not a single point, and does not contain either endpoint of any interval I_β with $\beta < \alpha$. We can do this because the set of endpoints to be avoided is countable and a Suslin line cannot be separable. Once this construction is complete, define a tree $\mathcal{T}$ whose vertices are the intervals I_α, ordered by reverse inclusion. This is a tree because $I_\alpha \supseteq I_\beta$ implies $\alpha \leq \beta$, so there cannot be any infinite decreasing sequences.

We verify that $\mathcal{T}$ is Suslin. The nonexistence of uncountable antichains follows from the fact that one cannot have uncountably many disjoint closed intervals in a Suslin line. Similarly, any branch corresponds to a nested sequence of closed intervals, and the gaps between the left endpoints of successive intervals in this sequence constitute a family of disjoint intervals. So if the line is Suslin there can be no uncountable branches. Finally, $\mathcal{T}$ has $\aleph_1$ vertices but, by the preceding, only countably many vertices at each level, so its height must be $\aleph_1$. We conclude that $\mathcal{T}$ is Suslin. $\qquad \square$

In fact every Suslin tree can be converted into a normal Suslin tree (Definition 18.1), but we do not need this result.

Theorem 24.5. *Assume MA + "$2^{\aleph_0} = \aleph_2$". Then Suslin lines do not exist.*

Proof. Assume a Suslin tree $\mathcal{T}$ exists. Let $\mathcal{T}'$ be the subtree consisting of all the vertices of $\mathcal{T}$ that have extensions at all higher levels. It is clear that T' still has no uncountable branches or antichains. Also, since each level of T is countable, if every vertex on some level had only countably many extensions then T could not have height $\aleph_1$; thus each level contains a vertex that belongs to T'. So T' is a Suslin tree. Moreover, for any vertex of T' the set of its extensions in T' must have height $\aleph_1$, for the same reason that T' must have height $\aleph_1$. This shows that every vertex of T' has extensions in T' at all higher levels.

For each $\alpha < \aleph_1$, let $D_\alpha \subseteq T'$ be the set of vertices of height at least α. Then each D_α is dense, and since T' is c.c.c. Martin's axiom implies that it has an ideal that meets every D_α. But this means it has an uncountable branch, a contradiction. We conclude that Suslin trees do not exist. By Lemma 24.4, it follows that Suslin lines do not exist. $\qquad\square$

Chapter 25

Whitehead's Problem, II*

We will now show that MA + "$2^{\aleph_0} = \aleph_2$" implies there is a counterexample to Whitehead's problem, i.e., an abelian group with only trivial extensions by $\mathbb{Z}$ that is not free. This result, which complements Theorem 21.5, is also due to Shelah. Together they show that the answer to Whitehead's problem for groups of cardinality $\aleph_1$ is independent of ZFC (assuming ZFC is consistent).

All we need is a transfinite strictly increasing sequence of countable free abelian groups $\mathcal{A}_\alpha$, $\alpha < \aleph_1$, with the properties that (1) $\mathcal{A}_0 = \{0\}$; (2) $\mathcal{A}_\alpha/\mathcal{A}_\beta$ is free for all $\beta < \alpha < \aleph_1$; and (3) $\mathcal{A} = \bigcup \mathcal{A}_\alpha$ is not free. We begin by showing how such a sequence can be constructed.

As we noted in Chapter 21, every extension of a free abelian group by $\mathbb{Z}$ splits. In fact, every extension of a free abelian group by any abelian group splits: if $\pi : \mathcal{B} \to \mathcal{A}$ is a surjective homomorphism and S is a basis for $\mathcal{A}$, choose $\rho(a)$ arbitrarily in the set $\pi^{-1}(a)$ for each $a \in S$, and extend ρ to $\mathcal{A}$ linearly. Then $\pi \circ \rho = \mathrm{id}_{\mathcal{A}}$ and we can decompose $\mathcal{B}$ as $\mathcal{B} = \rho(\mathcal{A}) \oplus \ker \pi$. In particular, if $\mathcal{C} \subseteq \mathcal{B}$ and $\mathcal{C}$ and $\mathcal{B}/\mathcal{C}$ are both free, then so is $\mathcal{B} \cong \mathcal{B}/\mathcal{C} \oplus \mathcal{C}$.

Lemma 25.1. *There exists a transfinite strictly increasing sequence of countable abelian groups $\mathcal{A}_\alpha$, $\alpha < \aleph_1$, with the properties*

(i) $\mathcal{A}_0 = \{0\}$
(ii) $\mathcal{A}_\alpha/\mathcal{A}_\beta$ is free for all $\beta < \alpha < \aleph_1$
(iii) $\mathcal{A} = \bigcup \mathcal{A}_\alpha$ is not free.

Proof. We describe the construction. Start with $\mathcal{A}_0 = \{0\}$. At successor stages, let $\mathcal{A}_{\alpha+1} = \mathcal{A}_\alpha \oplus \mathbb{Z}$. Assuming property (ii) held previously, it will still hold because $\mathcal{A}_{\alpha+1}/\mathcal{A}_\beta \cong \mathcal{A}_\alpha/\mathcal{A}_\beta \oplus \mathbb{Z}$ for $\beta \le \alpha$. At limit stages, choose a sequence (α_n) with $\alpha_0 = 0$ that strictly increases to α. Use the fact that $\mathcal{A}_{\alpha_{n+1}}/\mathcal{A}_{\alpha_n}$ is free for all n (by the induction hypothesis) to write $\mathcal{A}_{<\alpha} \cong \bigoplus \mathcal{A}_{\alpha_{n+1}}/\mathcal{A}_{\alpha_n}$, and then find a basis $S = \bigcup S_n$ of $\mathcal{A}_{<\alpha}$ where each S_n corresponds to a basis of $\mathcal{A}_{\alpha_n+1}/\mathcal{A}_{\alpha_n}$ — that is, where $\{a + \mathcal{A}_{\alpha_n} \mid a \in S_n\}$ is a basis of $\mathcal{A}_{\alpha_n+1}/\mathcal{A}_{\alpha_n}$ for all n. Then choose a single element $a_n \in S_n$ for each n and let $\mathcal{A}_\alpha$ be the group generated by $\mathcal{A}_{<\alpha}$ and the additional generators $b_0, b_1, \ldots$ subject to the relations

$$b_n - 2b_{n+1} = a_n$$

$(n = 0, 1, \ldots)$.

We prove that property (ii) is preserved at limit stages. First, observe that for all n the quotient $\mathcal{A}_\alpha/\mathcal{A}_{\alpha_n}$ is free with basis

$$\left(\bigcup_{k=n}^{\infty} S_k \setminus \{a_k\} \right) \cup \{b_n, b_{n+1}, \ldots\}.$$

Thus this quotient is free. But then for any $\beta < \alpha$ we have $\beta < \alpha_n$ for some n, and $\mathcal{A}_\alpha \cong \mathcal{A}_\alpha/\mathcal{A}_{\alpha_n} \oplus \mathcal{A}_{\alpha_n}$ implies that

$$\mathcal{A}_\alpha/\mathcal{A}_\beta \cong \mathcal{A}_\alpha/\mathcal{A}_{\alpha_n} \oplus \mathcal{A}_{\alpha_n}/\mathcal{A}_\beta,$$

which is a direct sum of free abelian groups and hence is free. So property (ii) still holds.

Finally, we must show that $\mathcal{A} = \bigcup_{\alpha < \aleph_1} \mathcal{A}_\alpha$ is not free. Suppose it is free and let S be a basis. Recursively define an increasing sequence of countable ordinals α_n as follows. Let $\alpha_0 = 0$. Given α_n, $\mathcal{A}_{\alpha_n}$ is a countable subgroup of $\mathcal{A}$ so it is contained in the span of a countable subset S_n of S. Find $\alpha_{n+1} > \alpha_n$ such that $\mathcal{A}_{\alpha_{n+1}}$ contains S_n.

Let $\alpha = \sup \alpha_n$. Then $\mathcal{A}_{<\alpha}$ is spanned by $\mathcal{A}_{<\alpha} \cap S$, so $\mathcal{A} = \mathcal{A}_{<\alpha} \oplus \mathcal{A}/\mathcal{A}_{<\alpha}$ and both summands are free. However, this is impossible because the new generators $b_0, b_1, \ldots$ which appear in $\mathcal{A}_\alpha$ satisfy

$$b_0 + \mathcal{A}_{<\alpha} = 2b_1 + \mathcal{A}_{<\alpha} = 4b_2 + \mathcal{A}_{<\alpha} = \cdots,$$

contradicting the freeness of $\mathcal{A}/\mathcal{A}_{<\alpha}$. So $\mathcal{A}$ cannot be free. $\square$

In particular, $\mathcal{A}_\alpha \cong \mathcal{A}_\alpha/\mathcal{A}_0$ is free for all α.

A subgroup $\mathcal{B}$ of a torsion-free group is *pure* if $na \in \mathcal{B}$ implies $a \in \mathcal{B}$, for any $n \geq 1$. Any countable subgroup $\mathcal{B}$ can be enlarged to a countable pure subgroup, namely the set of a such that $na \in \mathcal{B}$ for some $n \geq 1$. This set is countable since lack of torsion implies that $na = nb$ for $n \geq 1$ implies $a = b$. Now let $\mathcal{A}$ be as in Lemma 25.1 and suppose we are given an abelian group $\mathcal{B}$ and a surjective homomorphism $\pi : \mathcal{B} \to \mathcal{A}$ with kernel $\mathbb{Z}$. Define $\mathcal{P}$ to be the set of all homomorphisms $\rho : \mathcal{A}' \to \mathcal{B}$ such that $\mathcal{A}'$ is a finitely generated pure subgroup of $\mathcal{A}$ and $\pi \circ \rho = \mathrm{id}_{\mathcal{A}'}$. That is, ρ is a splitting of $\pi|_{\pi^{-1}(\mathcal{A}')}$. Order $\mathcal{P}$ by inclusion, i.e., set $\rho_1 \leq \rho_2$ if ρ_2 extends ρ_1.

In the proof of the next result we invoke a standard fact which states that any subgroup of a free abelian group is free.

Lemma 25.2. $\mathcal{P}$ *is c.c.c.*

Proof. Let $Z \subseteq \mathcal{P}$ be uncountable. We claim that there is a pure free subgroup $\mathcal{C}$ of $\mathcal{A}$ which contains the domains of uncountably many $\rho \in Z$. Assuming this claim, let S be a basis for $\mathcal{C}$; then the domain of any $\rho \in Z$, if contained in $\mathcal{C}$, is contained in the span of a finite subset S_ρ of S. By the Δ-systems lemma (Lemma 13.4) there is then an uncountable subset $Z' \subseteq Z$ and a finite set $S_0 \subseteq S$ such that $S_\rho \cap S_{\rho'} = S_0$ for all distinct $\rho, \rho' \in Z'$. For each $\rho \in Z'$ let $\tilde{\rho} \in \mathcal{P}$ be an extension of ρ to the span of S_ρ; this is possible because the domain of ρ is pure, so that $\mathrm{span}(S_\rho)/\mathrm{dom}(\rho)$ has no torsion and hence is free by the structure theorem for finitely generated abelian groups. Then there are only countably many possible functions $\tilde{\rho}|_{S_0}$, so some distinct $\tilde{\rho}, \tilde{\rho}' \in Z'$ have the same restriction to S_0 and hence they have a common extension to the span of $S_\rho \cup S_{\rho'}$. Thus ρ and ρ' are compatible, and we conclude that P is c.c.c.

To prove the claim, find $n \in \mathbb{N}$ such that the domains of uncountably many $\rho \in Z$ have bases of size n. Let Z_n be the set of all such ρ. Then let $\mathcal{C}_0$ be a maximal pure subgroup of $\mathcal{A}$ having the property that $\mathcal{C}_0 \subseteq \mathrm{dom}(\rho)$ for uncountably many $\rho \in Z_n$. Let Z' be the set of all $\rho \in Z_n$ whose domain contains $\mathcal{C}_0$.

Recursively define a transfinite sequence of homomorphisms $\rho_\alpha \in Z'$ and countable pure subgroups $\mathcal{C}_\alpha$ of $\mathcal{A}$ for $\alpha < \aleph_1$ as follows. Suppose $\mathcal{C}_\beta$ has been defined for all $\beta < \alpha$ and find α' such that $\mathcal{C}_{<\alpha} \subseteq \mathcal{A}_{\alpha'}$. For all but countably many $\rho \in Z'$ we have $\mathrm{dom}(\rho) \cap \mathcal{A}_{\alpha'} = \mathcal{C}_0$; otherwise, since $\mathcal{A}_{\alpha'}$ is countable, some $a \in \mathcal{A}_{\alpha'} \setminus \mathcal{C}_0$ would be contained in the domains of uncountably many $\rho \in Z'$ and this would contradict maximality of $\mathcal{C}_0$. So choose $\rho_\alpha \in Z'$ distinct from all ρ_β for $\beta < \alpha$ and such that $\mathrm{dom}(\rho_\alpha) \cap \mathcal{A}_{\alpha'} = \mathcal{C}_0$. Then let $\mathcal{C}_\alpha$ be the smallest pure subgroup of $\mathcal{A}$ that contains both $\mathcal{C}_{<\alpha}$ and $\mathrm{dom}(\rho_\alpha)$. If $\mathcal{C}_\alpha \subseteq \mathcal{A}_{\beta'}$ then $\mathcal{C}_\alpha/\mathcal{C}_{<\alpha}$ is isomorphic to a subgroup of $\mathcal{A}_{\beta'}/\mathcal{A}_{\alpha'}$, which is free, so that $\mathcal{C}_\alpha \cong \mathcal{C}_\alpha/\mathcal{C}_{<\alpha} \oplus \mathcal{C}_{<\alpha}$ represents $\mathcal{C}_\alpha$ as a direct sum of free abelian groups.

Finally, let $\mathcal{C} = \bigcup_{\alpha < \aleph_1} \mathcal{C}_\alpha$. Then $\mathcal{C} \cong \bigoplus \mathcal{C}_\alpha/\mathcal{C}_{<\alpha}$ is a direct sum of free abelian groups so it is free, and it is pure because each $\mathcal{C}_\alpha$ is pure. It contains $\mathrm{dom}(\rho_\alpha)$ for all $\alpha < \aleph_1$, so the claim is proven. $\square$

Theorem 25.3. *Assume MA + "$2^{\aleph_0} = \aleph_2$". Then there is a group of cardinality $\aleph_1$ that is not free but all of whose extensions by $\mathbb{Z}$ are trivial.*

Proof.   Let $\mathcal{A}$ be as in Lemma 25.1, let $\pi : \mathcal{B} \to \mathcal{A}$ be an extension by $\mathbb{Z}$, and let $\mathcal{P}$ be the poset defined just before Lemma 25.2. We must show that π splits.

For each $a \in \mathcal{A}$ let D_a be the set of all $\rho \in \mathcal{P}$ whose domain contains a. We claim that D_a is dense. To see this, let $\rho \in \mathcal{P}$ and let $\mathcal{A}_\alpha$ contain both a and the domain of ρ. Let S be a basis for $\mathcal{A}_\alpha$ and let S_0 be a finite subset of S whose span contains a and the domain of ρ. Let $\mathcal{A}'$ be the span of S_0; then $\mathcal{A}'/\mathrm{dom}(\rho)$ is free (for the same reason $\mathrm{span}(S_\rho)/\mathrm{dom}(\rho)$ was free in the proof of Lemma 25.2), so we can choose a basis and extend ρ to a splitting of $\pi|_{\pi^{-1}(\mathcal{A}')} : \pi^{-1}(\mathcal{A}') \to \mathcal{A}'$ one basis element at a time. Also, $\mathcal{A}'$ is clearly pure within $\mathcal{A}_\alpha$, so it is pure in $\mathcal{A}_{\alpha'} \cong \mathcal{A}_{\alpha'}/\mathcal{A}_\alpha \oplus \mathcal{A}_\alpha$ for all $\alpha' \geq \alpha$, and therefore it is pure in $\mathcal{A}$. This shows that ρ has an extension in D_a.

Martin's axiom now implies that there is an ideal G of $\mathcal{P}$ that intersects every D_a. The union of G is then a homomorphism from $\mathcal{A}$ to $\mathcal{B}$ that splits π, as desired.   □

Chapter 26

The Open Coloring Axiom

Let $\mathcal{X}$ be a separable metric space. An *open coloring* of $\mathcal{X}$ is a symmetric open subset K of $\mathcal{X}^2 \setminus \Delta_\mathcal{X}$ where $\Delta_\mathcal{X} = \{\langle x, x \rangle \mid x \in \mathcal{X}\}$. A subset A of $\mathcal{X}$ is *0-homogeneous* if $\langle x, y \rangle \in K$ for all distinct $x, y \in A$ and *1-homogeneous* if $\langle x, y \rangle \notin K$ for all distinct $x, y \in A$. The *open coloring axiom (OCA)* says that if $\mathcal{X}$ is any separable metric space and K is any open coloring of $\mathcal{X}$, then either $\mathcal{X}$ has an uncountable 0-homogeneous set or $\mathcal{X}$ can be covered by countably many 1-homogeneous sets.

We think of K as describing a 2-coloring of the edges of a complete graph on $\mathcal{X}$. Openness means that if the edge from x to y has the color 0 then so do the edges from x' to y' for all x' and y' sufficiently close to x and y. Thus, the closure of any 1-homogeneous set is 1-homogeneous.

The key lemma which enables us to force OCA is the following. This lemma is provable in ZFC.

Lemma 26.1. *Assume CH and let K be an open coloring of a separable metric space $\mathcal{X}$. If $\mathcal{X}$ cannot be covered by countably many 1-homogeneous sets, then there is an uncountable subset A of $\mathcal{X}$ such that the family P of finite 0-homogeneous subsets of A is c.c.c.*

Proof. For $n \in \mathbb{N}$, $U \subseteq \mathcal{X}^n$ open, and $\vec{x} = \langle x_1, \ldots, x_n \rangle \in U$, let $U_{\vec{x}}$ be the set of $\vec{y} \in U$ such that $\langle x_i, y_i \rangle \in K$ for $1 \leq i \leq n$. Also, for any partial function f from $\mathcal{X}^n$ to $\mathcal{X}$ and any $\vec{x} \in \mathcal{X}^n$, let $\phi_f(\vec{x})$ be the intersection of the sets $\overline{f(U_{\vec{x}})}$ as U ranges over the open subsets of $\mathcal{X}^n$ which contain $\vec{x}$ (interpreting $f(U_{\vec{x}})$ as $f(U_{\vec{x}} \cap \mathrm{dom}(f))$).

Enumerate the countable partial functions (i.e., with countable domains) from finite powers of $\mathcal{X}$ into $\mathcal{X}$ as $\{f_\beta \mid \beta < \aleph_1\}$ and enumerate the closed 1-homogeneous subsets of $\mathcal{X}$ as $\{C_\beta \mid \beta < \aleph_1\}$. Then recursively construct a transfinite sequence $\{x_\alpha \mid \alpha < \aleph_1\}$ of points in $\mathcal{X}$ such that, letting $A_\alpha = \{ x_\beta \mid \beta < \alpha \}$, we have

(i) $x_\alpha \notin A_\alpha$
(ii) $x_\alpha \notin \bigcup_{\beta < \alpha} C_\beta$
(iii) $x_\alpha \notin \phi_{f_\beta}(\vec{x})$ for any $\beta < \alpha$ and any $\vec{x} \in A_\alpha^n \cap \mathrm{dom}(f_\beta)$ (where $f_\beta :$ $\mathcal{X}^n \to \mathcal{X}$) such that $\phi_{f_\beta}(\vec{x})$ is 1-homogeneous.

We can do this because $\mathcal{X}$ is not covered by any countable family of 1-homogeneous sets. Let $A = \{ x_\alpha \mid \alpha < \aleph_1 \}$.

Let P be the family of finite 0-homogeneous subsets of A. Suppose we can show that any uncountable family of pairwise disjoint elements of P contains a pair of compatible elements. By the Δ-system lemma (Lemma 13.4), any uncountable $S \subseteq P$ contains an uncountable subset S' any two elements of which have the same intersection r. Applying the presumed result to the elements $p \setminus r$, for p in S', yields $p, q \in S'$ such that p is 0-homogeneous, q is 0-homogeneous, and $(p \cup q) \setminus r$ is 0-homogeneous, which together imply that $p \cup q$ is 0-homogeneous. So it will suffice to show that any uncountable family of pairwise disjoint elements of P contains a pair of compatible elements. Fix such a family S; we can assume all the elements of S have the same cardinality n. We prove the result by induction on n.

When $n = 1$, the elements of S are all singletons. If $\langle x, y \rangle \notin K$ for all distinct $\{x\}, \{y\} \in S$, then the closure of $\{ x \mid \{x\} \in S \}$ is 1-homogeneous and hence equals some C_β, but then we would have $\{x_\alpha\} \notin S$ for any $\alpha > \beta$, contradicting uncountability of S. So some distinct $\{x\}, \{y\} \in S$ must be compatible in P.

Now suppose $n > 1$. Enumerate S as $\{ p_\beta \mid \beta < \aleph_1 \}$, and for each β write $p_\beta = \{ x_{\alpha_{\beta,1}}, \ldots, x_{\alpha_{\beta,n}} \}$ where $\alpha_{\beta,1} < \cdots < \alpha_{\beta,n}$. Fix a countable basis $\mathcal{B}$ for the topology on $\mathcal{X}$ and let $\mathcal{B}^n = \{ U_1 \times \cdots \times U_n \mid U_i \in \mathcal{B}$ for $1 \le i \le n \}$ be the corresponding basis for $\mathcal{X}^n$. For each p_β we can find neighborhoods U_1 of $x_{\alpha_{\beta,1}}$, $\ldots$, U_n of $x_{\alpha_{\beta,n}}$ such that $U_i \times U_j \subseteq K$ for all $i \ne j$. Since $\mathcal{B}^n$ is countable, by replacing S with an uncountable subset we can arrange that a single $U_1 \times \cdots \times U_n$ contains every $\langle x_{\alpha_{\beta,1}}, \ldots, x_{\alpha_{\beta,n}} \rangle$. Then the pair $\langle x_{\alpha_{\beta,i}}, x_{\alpha_{\gamma,j}} \rangle$ belongs to K for every β, γ, and $i \ne j$. So we just have to find β and γ with $\langle x_{\alpha_{\beta,i}}, x_{\alpha_{\gamma,i}} \rangle \in K$ for all i.

For each β set $\vec{y}_\beta = \langle x_{\alpha_{\beta,1}}, \ldots, x_{\alpha_{\beta,n-1}} \rangle \in \mathcal{X}^{n-1}$ and $z_\beta = x_{\alpha_{\beta,n}}$ and define a partial function g from $\mathcal{X}^{n-1}$ to $\mathcal{X}$ by setting $g(\vec{y}_\beta) = z_\beta$ for each β. We claim that $z_\beta \in \phi_g(\vec{y}_\beta)$ for all but countably many β. For each β with $z_\beta \notin \phi_g(\vec{y}_\beta)$, find $U^\beta \in \mathcal{B}^{n-1}$ and $V^\beta \in \mathcal{B}$ such that $\vec{y}_\beta \in U^\beta$, $z_\beta \in V^\beta$, and V^β is disjoint from $g(U^\beta_{\vec{y}_\beta})$. Since there are only countably many possible choices for U^β and V^β, if the claim fails then some pair of U and V must be shared by uncountably many values of β. By the induction hypothesis we can then find distinct β, β' such that $U^\beta = U^{\beta'}$, $V^\beta = V^{\beta'}$, and the set $\{x_{\alpha_{\beta,1}}, \ldots, x_{\alpha_{\beta,n-1}}, x_{\alpha_{\beta',1}}, \ldots, x_{\alpha_{\beta',n-1}}\}$ is 0-homogeneous. But then $\vec{y}_{\beta'} \in U^\beta_{\vec{y}_\beta}$ and hence $g(U^\beta_{\vec{y}_\beta})$ intersects V^β, a contradiction. This proves the claim.

Now the graph of g is contained in the separable metric space $\mathcal{X}^n$, so it has a countable dense subset, which will be the graph of some f_{β_0}. By the claim we can find β such that $\alpha_{\beta,1} > \beta_0$ and $z_\beta \in \phi_g(\vec{y}_\beta) = \phi_{f_{\beta_0}}(\vec{y}_\beta)$. Since $\alpha_{\beta,1} > \beta_0$, the choice of $z_\beta = x_{\alpha_{\beta,n}}$ ensures that the set $\phi_{f_{\beta_0}}(\vec{y}_\beta)$ cannot be 1-homogeneous, so we can find distinct $x, x' \in \phi_{f_{\beta_0}}(\vec{y}_\beta)$ such that $\langle x, x' \rangle \in K$. Find open sets $U, U' \subseteq X$ such that $\langle x, x' \rangle \in U \times U' \subseteq K$. By the definition of $\phi_{f_{\beta_0}}$ there must exist γ such that $f_{\beta_0}(\vec{y}_\gamma) \in U$. Then $\{x_{\alpha_{\beta,1}}, \ldots, x_{\alpha_{\beta,n-1}}, x_{\alpha_{\gamma,1}}, \ldots, x_{\alpha_{\gamma,n-1}}\}$ is 0-homogeneous so we can find a neighborhood of $\langle x_{\alpha_{\beta,1}}, \ldots, x_{\alpha_{\beta,n-1}} \rangle$ every element of which has 0-homogeneous union with $\langle x_{\alpha_{\gamma,1}}, \ldots, x_{\alpha_{\gamma,n-1}} \rangle$, and we can choose γ' such that $\langle x_{\alpha_{\gamma',1}}, \ldots, x_{\alpha_{\gamma',n-1}} \rangle$ belongs to this neighborhood and $f_{\beta_0}(\vec{y}_{\gamma'}) \in U'$. Then $p_\gamma \cup p_{\gamma'}$ is 0-homogeneous, which completes the proof. $\square$

Theorem 26.2. *There is a forcing notion P such that $\mathbf{M}[G] \models OCA$ for any generic ideal G of P.*

The proof of Theorem 26.2 is somewhat complex but it does not introduce any new ideas, so we merely sketch it. We can use Lemma 26.1 to force an uncountable 0-homogeneous set whenever we have an open coloring for which $\mathcal{X}$ is not covered by countably many 1-homogeneous sets. Since we have to handle all possible open colorings of separable metric spaces, we will apply an iterated forcing construction. Lemma 26.1 requires CH, so we start by forcing CH and then iterate along $\aleph_2$ stages, ensuring that each P_β is c.c.c. and has cardinality $\aleph_1$. CH is maintained throughout the construction since any subset of $\mathbb{N}$ in a forcing extension is the value of a countably-generated name as in Lemma 20.6, and there are only $\aleph_1^{\aleph_0} = \aleph_1$ of these. CH will fail at the final stage $\alpha = \aleph_2$, but not before.

Let a *v-space* be a separable metric space equipped with an open coloring with respect to which it is not covered by any family of countably many 1-homogeneous sets. An extra complication that appears in this argument, as opposed to the proof of Theorem 23.3, is that new v-spaces could appear in the final forcing extension $\mathbf{M}[G]$ that were not available in any earlier extension $\mathbf{M}[G_\beta]$. Moreover, since $2^{\aleph_0}$ finally settles on the value $\aleph_2$, there are too many v-spaces to handle separately in only $\aleph_2$ steps. So we have to deal with v-spaces that appear in $\mathbf{M}[G]$ by some kind of approximation. This is accomplished as follows. We only need to consider v-spaces in $\mathbf{M}[G]$ whose underlying set X is a subset of $\aleph_2$ and such that the initial ω is a dense subset. Then any metric on X is a function from $X^2 \subseteq \aleph_2^2$ into $\mathbb{R}$, and any open coloring of X is the union of countably many balls about points in a dense subset and hence is coded by a function from a subset of ω into $(0, \infty)$. So there will be a name for X which is generated by a set consisting of elements of the form $\langle \check{\alpha}, p \rangle$, with each $\alpha \in \aleph_2$ paired with countably many $p \in P^*$, and similarly for the other data.

We can further regularize the relevant data by arranging that the final forcing notion P^* is in bijection with $\aleph_2$ in such a way that each P_β is in bijection with an initial segment of $\aleph_2$. We just build up the desired bijection as we build up the iterated forcing construction. The result is that any v-space that appears in $\mathbf{M}[G]$ is the value of a name which is coded by data that is available throughout the construction in $\mathbf{M}$. The key point is that as the construction proceeds, on a club set of levels β the restriction of this data to P_β will evaluate to a v-space in the corresponding extension $\mathbf{M}[G_\beta]$.

So if we use a version of the diamond principle for trees of height $\aleph_2$ to choose a forcing poset at each level, then for every v-space in $\mathbf{M}[G]$ there will be a stage at which a name for it restricts to a name which evaluates to a v-space in $\mathbf{M}[G_\beta]$ that is chosen by diamond. Then an uncountable 0-homogeneous set in the latter v-space will appear at the next stage, and hence the former v-space (which contains the latter) must also have an uncountable 0-homogeneous set.

To summarize, we first force CH, then a version of diamond for the complete $\aleph_2$-$\aleph_2$ tree, and we then proceed to build a finite support iterated forcing construction using diamond to destroy any possible counterexample to OCA.

Metatheorem 26.3. *If ZFC is consistent, then so is ZFC + OCA.*

We can achieve a little more by interleaving the iterated forcing constructions of Theorems 23.3 and 26.3. As long as the OCA forcing takes place on a club set our use of diamond will still succeed, and it does not matter exactly which levels we force MA on as long as every c.c.c. ordering of $\aleph_1$ gets taken care of eventually. This produces the following result.

Metatheorem 26.4. *If ZFC is consistent, then so is ZFC + OCA + MA.*

Chapter 27

Self-Homeomorphisms of $\beta\mathbb{N} \setminus \mathbb{N}$, II*

OCA + MA implies that every self-homeomorphism of $\beta\mathbb{N} \setminus \mathbb{N}$ arises from an almost permutation of $\mathbb{N}$. This result, which complements Theorem 15.3, is due to Veličković. Together they show that the triviality of all self-homeomorphisms of $\beta\mathbb{N} \setminus \mathbb{N}$ is independent of ZFC (assuming ZFC is consistent).

Before we discuss Veličković's result, we present two simpler applications of OCA. In these simpler applications we do not need the full dichotomy asserted by OCA, only the weaker assertion, under the hypothesis that $\mathcal{X}$ is uncountable, that $\mathcal{X}$ has either an uncountable 0-homogeneous subset or an uncountable 1-homogeneous subset.

First, we have an elementary result about monotonic functions.

Theorem 27.1. *Assume OCA and let $A \subseteq \mathbb{R}$ be uncountable. Then every function from A to $\mathbb{R}$ is monotonic on an uncountable set.*

Proof. Fix $f : A \to \mathbb{R}$. Its graph is contained in $\mathbb{R}^2$ and so inherits a separable metric structure. Define an open coloring K on the graph of f by letting the pair of points $\langle x, f(x) \rangle$ and $\langle x', f(x') \rangle$ belong to K if the slope of the line joining them is strictly positive. By OCA there is either an uncountable 0-homogeneous set or an uncountable 1-homogeneous set, in either case the graph of a monotone function. $\qquad\square$

Next we consider the *Baire space* $\mathbb{N}^{\mathbb{N}}$ of functions from $\mathbb{N}$ to itself. Topologically it is the product of countably many copies of the discrete space $\mathbb{N}$. According to a standard theorem from topology, any product of countably many separable metrizable spaces is separable and metrizable. Given $f, g : \mathbb{N} \to \mathbb{N}$, write $f \leq_* g$ if g *eventually dominates* f, i.e., $f(n) \leq g(n)$ for all but finitely many n. Given any countable family (f_n), the function

$$g(n) = \max(f_1(n), \ldots, f_n(n))$$

eventually dominates each f_n; we say that the family (f_n) is *bounded*.

Theorem 27.2. *Assume OCA. Then every subset of $\mathbb{N}^{\mathbb{N}}$ of cardinality $\aleph_1$ is bounded.*

Proof. Let $\{f_\alpha \mid \alpha < \aleph_1\}$ be a family of functions in $\mathbb{N}^{\mathbb{N}}$ of cardinality $\aleph_1$. Recursively construct a sequence of functions f'_α, $\alpha < \aleph_1$, such that (1) each f'_α is strictly increasing; (2) $f_\alpha \leq_* f'_\alpha$; and (3) $f'_\beta \leq_* f'_\alpha$ for all $\beta < \alpha$. We will show that the set $X = \{f'_\alpha \mid \alpha < \aleph_1\}$ is bounded.

X inherits a separable metric from $\mathbb{N}^{\mathbb{N}}$. Define $K' \subset X^2 \setminus \Delta_X$ to be the set of pairs $\langle f, g \rangle$ such that either $f(n) \leq g(n)$ for all n or vice versa; this set is closed in the product topology on $\mathbb{N}^{\mathbb{N}}$, so its complement is an open coloring. By OCA, X has either an uncountable 0-homogeneous subset or an uncountable 1-homogeneous subset. But an uncountable 1-homogeneous subset of X is impossible: since X is well-ordered under eventual dominance and any two elements of the subset are strictly comparable, this would mean that we have an uncountable subset of $\mathbb{N}^{\mathbb{N}}$ that is well-ordered under strict comparability. For each function f in this subset find a point $(n, f(n))$ in its graph that does not lie in the graph of its successor; then distinct functions are assigned distinct points, but there are only countably many elements of $\mathbb{N}^2$, a contradiction.

Thus we must have an uncountable 0-homogeneous subset Y of X. It will suffice to show that Y is bounded, because every element of X lies below some element of Y. First, for each $n \in \mathbb{N}$ and each $\vec{k} = (k_1, \ldots, k_n) \in \mathbb{N}^n$, choose a function $f_{\vec{k}} \in Y$ which satisfies $f_{\vec{k}}(i) = k_i$ ($1 \leq i \leq n$) if one exists. There are only countably many such functions $f_{\vec{k}}$, so there exists $f_\gamma \in Y$ which eventually dominates them all.

There are still uncountably many $f \in Y$ which eventually dominate f_γ, so there exists n_1 such that $Z = \{f \in Y \mid f(n) \geq f_\gamma(n)$ for all $n \geq n_1\}$ is uncountable. Next, if $\{f(n) \mid f \in Z\}$ is finite for all n then Z is bounded, so Y is bounded too. Otherwise there exists n_2 such that $\{f(n_2) \mid f \in Z\}$ is infinite. Letting n_2 be minimal with this property, there is an upper bound on $f(n_2 - 1)$ for $f \in Z$, so some $\vec{k} = (k_1, \ldots, k_{n_2-1})$ must have infinitely many extensions to k_{n_2} which make it an initial segment of a function in Z. Finally, find $n_3 \geq \max(n_1, n_2)$ such that $f_\gamma(n) \geq f_{\vec{k}}(n)$ for all $n \geq n_3$ and find $f_\delta \in Z$ such that $\vec{k}$ is an initial segment of f_δ and $f_\delta(n_2) \geq f_{\vec{k}}(n_3)$.

We now have $f_\delta(n) = f_{\vec{k}}(n)$ for $1 \leq n < n_2$, $f_\delta(n) \geq f_\delta(n_2) \geq f_{\vec{k}}(n_3) > f_{\vec{k}}(n)$ for $n_2 \leq n < n_3$, and $f_\delta(n) \geq f_\gamma(n) \geq f_{\vec{k}}(n)$ for all $n \geq n_3$. This contradicts 0-homogeneity of Y. So Z must have been bounded. $\qquad\square$

Incidentally, this result shows that OCA implies $\neg$ CH.

Now we turn to self-homeomorphisms of $\beta\mathbb{N} \setminus \mathbb{N}$. As in Chapter 15, we equivalently work with order-automorphisms of $\mathcal{P}(\mathbb{N})/\mathrm{fin}$. Now if $A \subseteq \mathbb{N}$ is infinite then any function $f : A \to \mathbb{N}$ induces a map from $\mathcal{P}(A)/\mathrm{fin}$ to $\mathcal{P}(\mathbb{N})/\mathrm{fin}$. We say that an order-automorphism ϕ of $\mathcal{P}(\mathbb{N})/\mathrm{fin}$ is *trivial on A* if its restriction to $\mathcal{P}(A)/\mathrm{fin}$ arises in this way. The full proof that OCA implies all order-automorphisms are trivial is too long to include here; instead, we present a key step which uses OCA to show that ϕ is trivial on some infinite $A \subseteq \mathbb{N}$. This is done using the following fact whose proof we omit. Give $\mathcal{P}(\mathbb{N})$ a topology by identifying it with the Cantor set.

Theorem 27.3. *Let ϕ be an order-automorphism of $\mathcal{P}(\mathbb{N})/\mathrm{fin}$ and suppose there is an infinite sequence of Borel functions $\psi_n : \mathcal{P}(\mathbb{N}) \to \mathcal{P}(\mathbb{N})$ such that for every $A \subseteq \mathbb{N}$ we have $\phi([A]) = [\psi_n(A)]$ for some n. Then ϕ is trivial.*

A family F of subsets of $\mathbb{N}$ is *almost disjoint* if $A \cap B$ is finite for all distinct $A, B \in F$. For instance, if we identify $\mathbb{N}$ with the vertices of an infinite binary tree, then the branches of the tree constitute a family of $2^{\aleph_0}$ almost disjoint subsets of $\mathbb{N}$. In general we say that an almost disjoint family F of subsets of $\mathbb{N}$ is *tree-like* if $\mathbb{N}$ can be identified with the vertices of an infinite binary tree in such a way that each $A \in F$ lies within a single branch.

Theorem 27.4. *Assume OCA. Let ϕ be an order-automorphism of $\mathcal{P}(\mathbb{N})/\mathrm{fin}$ and let F be a tree-like almost disjoint family of subsets of $\mathbb{N}$. Then ϕ is trivial on all but countably many $A \in F$.*

Proof. Let $\mathcal{X} \subset \mathcal{P}(\mathbb{N})^2$ be the set of all pairs $\langle A, A' \rangle$ of infinite subsets of $\mathbb{N}$ such that $A \subseteq A' \subseteq C$ for some $C \in F$. Identify $\mathbb{N}$ with the vertices of an infinite binary tree such that each $C \in F$ lies in a single branch, fix a lifting $\psi : \mathcal{P}(\mathbb{N}) \to \mathcal{P}(\mathbb{N})$ of ϕ (i.e., ψ satisfies $[\psi(A)] = \phi([A])$ for all $A \subseteq \mathbb{N}$), and define $K \subset \mathcal{X}^2 \setminus \Delta_{\mathcal{X}}$ by letting $\langle \langle A, A' \rangle, \langle B, B' \rangle \rangle \in K$ if

(i) A' and B' lie in different branches
(ii) $A \cap B' = A' \cap B$
(iii) $\psi(A) \cap \psi(B') \neq \psi(A') \cap \psi(B)$.

K is open for the topology $\mathcal{X}$ inherits when embedded in $\mathcal{P}(\mathbb{N})^4$ via the map $\langle A, A' \rangle \mapsto \langle A, A', \psi(A), \psi(A') \rangle$. Condition (ii) is maintained in a neighborhood since the branches containing A' and B' diverge by condition (i).

We claim that $\mathcal{X}$ has no uncountable 0-homogeneous subsets. To see this, suppose Y is an uncountable 0-homogeneous set and let

$$D = \bigcup \{A \subseteq \mathbb{N} \mid \langle A, A' \rangle \in Y \text{ for some } A'\}.$$

Then for any $\langle A, A' \rangle \in Y$ we have $D \cap A' = A$, and therefore $\psi(D) \cap \psi(A')$ and $\psi(A)$ have finite symmetric difference. So we can find $n \in \mathbb{N}$ and an uncountable subset $Z \subseteq Y$ such that $(\psi(D) \cap \psi(A')) \triangle \psi(A) \subseteq n$ and $\psi(A) \subseteq \psi(A') \cup n$ for all $\langle A, A' \rangle \in Z$. Thus there must be distinct $\langle A, A' \rangle$ and $\langle B, B' \rangle$ in Z such that $\psi(A) \cap n = \psi(B) \cap n$ and $\psi(A') \cap n = \psi(B') \cap n$, which implies that $\psi(A) \cap \psi(B') = \psi(A') \cap \psi(B)$, contradicting condition (iii) in the definition of K. This proves the claim.

OCA now implies that there is a countable cover (X_n) of $\mathcal{X}$ by 1-homogeneous sets. For each n let Y_n be a countable dense subset of X_n, and define

$$F_0 = \{C \in F \mid A \subseteq A' \subseteq C \text{ for some } \langle A, A' \rangle \text{ in some } Y_n\}.$$

This set is countable because each $\langle A, A' \rangle$ determines a unique C by almost disjointness. We will show that ϕ is trivial on every $C \in F \setminus F_0$.

Let $C \in F \setminus F_0$. We now claim that we can partition C into disjoint subsets C_1 and C_2 such that for any $\langle A, A' \rangle$ that belongs to some X_n with $A' \subseteq C_1$ or $A' \subseteq C_2$, for all $m \in \mathbb{N}$ there exists $\langle B, B' \rangle \in Y_n$ such that

(i) $A \cap B' = A' \cap B$
(ii) $A \cap m = B \cap m$ and $A' \cap m = B' \cap m$
(iii) $\psi(A) \cap m = \psi(B) \cap m$ and $\psi(A') \cap m = \psi(B') \cap m$.

To do this, construct an increasing infinite sequence (n_k) as follows. Start with $n_0 = 0$. Given n_k, choose $n_{k+1} > n_k$ sufficiently large that for every $i \leq k$ and $\langle A, A' \rangle \in X_i$, there exists $\langle B, B' \rangle \in Y_i$ such that $A \cap n_k = B \cap n_k$, $A' \cap n_k = B' \cap n_k$, $\psi(A) \cap n_k = \psi(B) \cap n_k$, $\psi(A') \cap n_k = \psi(B') \cap n_k$, and $B' \cap C \subseteq n_{k+1}$. The first four conditions are easy since Y_i is dense in X_i, and the last is possible since each X_i is 1-homogeneous (proof by examining cases). Now let

$$C_1 = C \cap ([n_0, n_1) \cup [n_2, n_3) \cup \cdots)$$

and

$$C_2 = C \cap ([n_1, n_2) \cup [n_3, n_4) \cup \cdots);$$

these verify the claim.

Finally, for each $n \in \mathbb{N}$ define a Borel function $\psi_n : \mathcal{P}(C_1) \to \mathcal{P}(\mathbb{N})$ by

$$\psi_n(B) = \bigcup \{\psi(C_1) \cap \psi(A) \mid \langle A, A' \rangle \in Y_n \text{ and } B \cap A' = C_1 \cap A\}.$$

Then $\psi(B) \triangle \psi_n(B)$ is finite if $\langle B, C_1 \rangle \in X_n$, so ϕ is trivial on C_1 by Theorem 27.3, and it is trivial on C_2 similarly. So it is trivial on C. $\square$

Chapter 28

Automorphisms of the Calkin Algebra, I*

Let H be a separable infinite-dimensional complex Hilbert space. For the sake of concreteness we can take $H = l^2$, the space of square-summable complex sequences. Recall from Chapter 16 that $\mathcal{B}(l^2)$ is the space of bounded linear operators from l^2 to itself and $\mathcal{K}(l^2) \subset \mathcal{B}(l^2)$ is the space of compact operators, operators approximated in norm by finite rank operators. $\mathcal{K}(l^2)$ is a closed two-sided ideal of $\mathcal{B}(l^2)$, which implies that the quotient space $\mathcal{C}(l^2) = \mathcal{B}(l^2)/\mathcal{K}(l^2)$ is also a C*-algebra. This space is called the *Calkin algebra*. We write $\pi : \mathcal{B}(l^2) \to \mathcal{C}(l^2)$ for the quotient map.

An operator A on l^2 is *diagonal* if the standard basis vectors e_n are all eigenvectors, i.e., $Ae_n = a_n e_n$ for some complex sequence $\vec{a} = (a_n)$. Equivalently, it can be regarded as the *multiplication operator* $M_{\vec{a}}$ which acts on the l^2 sequence $\vec{b} = (b_n)$ by $M_{\vec{a}}\vec{b} = (a_n b_n)$. It is easy to see that the operator $M_{\vec{a}}$ is bounded if and only if the sequence $\vec{a}$ is bounded, so that the set of diagonal operators in $\mathcal{B}(l^2)$ is naturally isomorphic (indeed, isometrically isomorphic) to the space l^∞ of bounded complex sequences. Moreover, $M_{\vec{a}}$ is compact if and only if $a_n \to 0$, making the set of compact diagonal operators isomorphic to the space c_0 of complex sequences which converge to zero. In this sense the "diagonal part" of the Calkin algebra is isomorphic to l^∞/c_0, which can be identified with the space $C(\beta\mathbb{N} \setminus \mathbb{N})$ of continuous complex-valued functions on the Stone-Čech remainder $\beta\mathbb{N} \setminus \mathbb{N}$. For this reason the Calkin algebra is considered to be a noncommutative analog of $\beta\mathbb{N} \setminus \mathbb{N}$ in the same way the space of $n \times n$ complex matrices is a noncommutative analog of an n-element set.

The question we addressed in Chapters 15 and 27 about self-homeomorphisms of $\beta\mathbb{N} \setminus \mathbb{N}$ also has an analog for the Calkin algebra. An *automorphism* of $\mathcal{C}(l^2)$ is a linear isomorphism $\phi : \mathcal{C}(l^2) \cong \mathcal{C}(l^2)$ which respects products and adjoints. For example, if $U \in \mathcal{C}(l^2)$ is unitary, i.e., we have $U^*U = UU^* = 1$ where 1 is the unit element, then the map $X \mapsto U^*XU$ is an automorphism; these are called *inner automorphisms*. Is every automorphism of $\mathcal{C}(l^2)$ inner?

The solution to this problem resembles the solution for $\beta\mathbb{N} \setminus \mathbb{N}$. In this chapter we will prove a theorem of Phillips and the author which states that CH implies outer automorphisms (automorphisms that are not inner) exist. We will discuss the consistency of all automorphisms being inner in the next chapter.

Let a *block sequence* be a strictly increasing sequence $E = (n_k)$ of natural numbers. We write $\mathcal{D}_E$ for the set of all bounded block diagonal operators supported on the blocks $[n_k, n_{k+1}) \times [n_k, n_{k+1})$; that is, $A \in \mathcal{B}(l^2)$ belongs to $\mathcal{D}_E$ if $\langle Ae_i, e_j \rangle \neq 0$ implies $n_k \leq i, j < n_{k+1}$ for some k. Define $E^{even} = (n_0, n_2, n_4, \ldots)$ and $E^{odd} = (n_1, n_3, n_5, \ldots)$, and let $\mathcal{D}_E^+$ be the set of all operators of the form $A_1 + A_2 + B$ with $A_1 \in \mathcal{D}_{E^{even}}$, $A_2 \in \mathcal{D}_{E^{odd}}$, and $B \in \mathcal{K}(l^2)$. Finally, if $E = (n_k)$ and $\tilde{E} = (\tilde{n}_k)$ are block sequences, say that E *eventually contains* $\tilde{E}$ if for all but finitely many k we have $[\tilde{n}_k, \tilde{n}_{k+1}) \subseteq [n_{k'}, n_{k'+1})$ for some k'.

Lemma 28.1. *For any bounded operator $A \in \mathcal{B}(l^2)$ and any block sequence $\tilde{E}$ there exists a block sequence E which eventually contains $\tilde{E}$ such that $A \in \mathcal{D}_E^+$.*

Proof. For $S \subseteq \mathbb{N}$ let $P_S \in \mathcal{B}(l^2)$ be the orthogonal projection onto $l^2(S) \subseteq l^2(\mathbb{N})$. First, observe that for any $m \in \mathbb{N}$ and $\epsilon > 0$, compactness of the operators $AP_{[0,m)}$ and $P_{[0,m)}A$ implies that

$$\|P_{[n,\infty)}AP_{[0,m)}\|, \|P_{[0,m)}AP_{[n,\infty)}\| \leq \epsilon$$

for sufficiently large n. Now set $n_0 = 0$ and recursively find $n_{k+1} > n_k$ such that $n_{k+1} = \tilde{n}_{k'}$ for some k' (where $\tilde{E} = (\tilde{n}_k)$) and

$$\|P_{[n_{k+1},\infty)}AP_{[0,n_k)}\|, \|P_{[0,n_k)}AP_{[n_{k+1},\infty)}\| \leq 2^{-k}.$$

It is clear that E eventually contains $\tilde{E}$. We claim that the operators

$$A_1 = \sum_{k=0}^{\infty}(P_{[n_{2k},n_{2k+2})}AP_{[n_{2k},n_{2k+2})} - P_{[n_{2k+1},n_{2k+2})}AP_{[n_{2k+1},n_{2k+2})})$$

and

$$A_2 = \sum_{k=0}^{\infty}(P_{[n_{2k+1},n_{2k+3})}AP_{[n_{2k+1},n_{2k+3})} - P_{[n_{2k+2},n_{2k+3})}AP_{[n_{2k+2},n_{2k+3})})$$

verify that A is in $\mathcal{D}_E^+$. For we can decompose $P_{[n_k,\infty)}(A - A_1 - A_2)$ as

$$P_{[n_k,\infty)}AP_{[0,n_{k-1})} + \sum_{m=k}^{\infty} (P_{[n_{m+1},\infty)}AP_{[n_{m-1},n_m)} + P_{[n_m,n_{m+1})}AP_{[n_{m+2},\infty)})$$

(draw a picture), so that the preceding estimates yield

$$\|P_{[n_k,\infty)}(A - A_1 - A_2)\| \le 2^{-k+1} + \sum_{m=k}^{\infty}(2^{-m} + 2^{-m-1}) = 5 \cdot 2^{-k}.$$

Thus $A - A_1 - A_2$ is approximated by the finite rank operators $P_{[0,n_k)}(A - A_1 - A_2)$, and is therefore compact. $\qquad\square$

A multiplication operator $M_{\vec{u}}$ in $\mathcal{B}(l^2)$ is unitary if and only if $|u_i| = 1$ for all i, i.e., $\vec{u} \in \mathbb{T}^{\mathbb{N}}$ where $\mathbb{T} \subset \mathbb{C}$ is the unit circle. Say that $\vec{u}$ is *stable* if $|u_{i+1} - u_i| \to 0$ as $i \to \infty$ and say that $\vec{u}$ and $\vec{v}$ *align* on $E = (n_k)$ if $|u_i\bar{u}_{n_{k_i}} - v_i\bar{v}_{n_{k_i}}| \to 0$ as $i \to \infty$, where k_i is defined by the condition $n_{k_i} \le i \le n_{k_i+1}$.

Lemma 28.2. *Let E be a block sequence and suppose $\vec{u}, \vec{v} \in \mathbb{T}^{\mathbb{N}}$ are stable and align on E. Then $M_{\vec{u}}^* A M_{\vec{u}} - M_{\vec{v}}^* A M_{\vec{v}}$ is compact for any $A \in \mathcal{D}_E^+$.*

Proof. Suppose $A \in \mathcal{D}_{E^{even}}$. Define $u_i' = u_i\bar{u}_{n_{2k}}$ and $v_i' = v_i\bar{v}_{n_{2k}}$ for $n_{2k} \le i < n_{2k+2}$; then $M_{\vec{u}'}^* A M_{\vec{u}'} = M_{\vec{u}}^* A M_{\vec{u}}$ and $M_{\vec{v}'}^* A M_{\vec{v}'} = M_{\vec{v}}^* A M_{\vec{v}}$. Since $\vec{u}$ and $\vec{v}$ are both stable and align on E, it follows that they align on E^{even}, which is to say that $|u_i' - v_i'| \to 0$ as $i \to \infty$. Then $M_{\vec{u}'} - M_{\vec{v}'} = M_{\vec{u}' - \vec{v}'}$ is compact, so that

$$M_{\vec{u}'}^* A M_{\vec{u}'} - M_{\vec{v}'}^* A M_{\vec{v}'} = M_{\vec{u}'}^* A (M_{\vec{u}'} - M_{\vec{v}'}) + (M_{\vec{u}'}^* - M_{\vec{v}'}^*) A M_{\vec{v}'}$$

is also compact. The case where $A \in \mathcal{D}_{E^{odd}}$ is similar and the case where $A \in \mathcal{K}(l^2)$ is trivial. So the result holds for all elements of $\mathcal{D}_E^+$. $\qquad\square$

Lemma 28.3. *Let E be a block sequence and let $\vec{u} \in \mathbb{T}^{\mathbb{N}}$ be stable. Then there is a stable $\vec{v} \in \mathbb{T}^{\mathbb{N}}$ that aligns with $\vec{u}$ on E but such that $M_{\vec{u}}^* B M_{\vec{u}} - M_{\vec{v}}^* B M_{\vec{v}}$ is not compact for some $B \in \mathcal{B}(l^2)$.*

Proof. Say $E = (n_k)$. Fix a sequence (z_k) of complex numbers of modulus 1 such that $|z_{k+1} - z_k| \to 0$ as $k \to \infty$ but for every N there exist $i, j > N$ with $|z_j - z_i| \ge 1$. For instance, $z_k = e^{2\phi i\sqrt{k}}$ works. Now define $v_i = z_k u_i$ for $n_k \le i < n_{k+1}$. It is clear that $\vec{v}$ is stable and aligns with $\vec{u}$ on E.

Find a pair of sequences of natural numbers (i_m) and (j_m) such that $i_0 < j_0 < i_1 < j_1 < \cdots$ and $|z_{j_m} - z_{i_m}| \ge 1$ for all m. Then define

$B \in \mathcal{B}(l^2)$ by $Be_{n_{i_m}} = e_{n_{j_m}}$ for all m and $Be_n = 0$ for all other e_n. For all m we then have

$$(M_{\vec{u}}^* BM_{\vec{u}} - M_{\vec{v}}^* BM_{\vec{v}})e_{n_{i_m}} = (1 - \bar{z}_{j_m} z_{i_m})\bar{u}_{n_{j_m}} u_{n_{i_m}} e_{n_{j_m}} = \lambda e_{n_{j_m}}$$

where $|\lambda| = |1 - \bar{z}_{j_m} z_{i_m}| = |z_{j_m} - z_{i_m}| \geq 1$. This shows that $M_{\vec{u}}^* BM_{\vec{u}} - M_{\vec{v}}^* BM_{\vec{v}}$ is not compact. $\qquad\Box$

Theorem 28.4. *Assume CH. Then the Calkin algebra has outer automorphisms.*

Proof. Enumerate the elements of $\mathcal{B}(l^2)$ as $\{A_\alpha \mid \alpha < \aleph_1\}$. We recursively define E_α and $\vec{u}_\alpha$, $\alpha < \aleph_1$, with the properties

(i) if $\beta \leq \alpha$ then E_α eventually contains E_β
(ii) $\vec{u}_\alpha$ is stable and aligns with $\vec{u}_\beta$ on E_β for all $\beta \leq \alpha$
(iii) $A_\alpha \in \mathcal{D}_{E_{\alpha+1}}^+$
(iv) $M_{\vec{u}_{\alpha+1}}^* BM_{\vec{u}_{\alpha+1}} - A_\alpha^* BA_\alpha$ is not compact for some $B \in \mathcal{D}_{E_{\alpha+1}}^+$

for all α. Start by letting $\vec{u}_0$ and E_0 be the sequences $(1, 1, 1, \ldots)$ and $(0, 1, 2, \ldots)$. At successor stages use Lemma 28.3 to find $\vec{v}$ that is stable and aligns with $\vec{u}_\alpha$ on E_α but such that $M_{\vec{u}_\alpha}^* BM_{\vec{u}_\alpha} - M_{\vec{v}}^* BM_{\vec{v}}$ is not compact for some $B \in \mathcal{B}(l^2)$. Since E_α eventually contains E_β and $\vec{u}_\alpha$ aligns with $\vec{u}_\beta$ on E_β for all $\beta \leq \alpha$, it follows that $\vec{v}$ will also align with $\vec{u}_\beta$ on E_β for all $\beta \leq \alpha$. Now set either $\vec{u}_{\alpha+1} = \vec{u}_\alpha$ or $\vec{u}_{\alpha+1} = \vec{v}$, whichever one makes $M_{\vec{u}_{\alpha+1}}^* BM_{\vec{u}_{\alpha+1}} - A_\alpha^* BA_\alpha$ not compact. Then use Lemma 28.1 to find $E_{\alpha+1}$ which eventually contains E_α and such that $A_\alpha, B \in \mathcal{D}_{E_{\alpha+1}}^+$. At limit stages we only need to meet conditions (i) and (ii); this is done by choosing a sequence (α_n) that increases to α and diagonalizing the corresponding sequences (u_{α_n}) and (E_{α_n}). The details are slightly tedious to write out but they are not difficult.

If E_α eventually contains E_β then $\mathcal{D}_{E_\beta}^+ \subseteq \mathcal{D}_{E_\alpha}^+$, so the sequence $(\mathcal{D}_{E_\alpha}^+)$ is increasing. For each α let $\mathcal{A}_\alpha = \pi(\mathcal{D}_{E_\alpha}^+)$ be the image of $\mathcal{D}_{E_\alpha}^+$ in the Calkin algebra and let ϕ_α be the inner automorphism of $\mathcal{C}(l^2)$ given by conjugating by $\pi(M_{\vec{u}_\alpha}) \in \mathcal{C}(l^2)$. By condition (ii) and Lemma 28.2, ϕ_α and ϕ_β agree on $\mathcal{A}_\beta$ whenever $\beta \leq \alpha$. So the sequence (ϕ_α) converges pointwise to an automorphism ϕ of $\mathcal{C}(l^2)$, and ϕ is not inner by condition (iv). $\qquad\Box$

As in the case of Theorem 15.3, with careful bookkeeping the proof of Theorem 28.4 can be used to show that under CH there are $2^{\aleph_1}$ distinct automorphisms of $\mathcal{C}(l^2)$, as opposed to only $\aleph_1$ inner automorphisms.

Chapter 29

Automorphisms of the Calkin Algebra, II*

We showed in the last chapter that CH implies the Calkin algebra has outer automorphisms. Conversely, Farah proved that OCA implies all automorphisms of the Calkin algebra are inner. Thus, the existence of outer automorphisms is independent of ZFC (assuming ZFC is consistent).

The proof that all automorphisms of $\mathcal{C}(l^2)$ are inner follows the same broad pattern as the proof that all self-homeomorphisms of $\beta\mathbb{N}\setminus\mathbb{N}$ are trivial. First one shows that every automorphism that has a lift to $\mathcal{B}(l^2)$ with some measurability property is inner (analogous to Theorem 27.3). Next one uses OCA and the preceding result to show that every automorphism is "locally" inner (analogous to Theorem 27.4). Finally, one uses OCA again to turn this local conclusion into a global one. In the $\beta\mathbb{N}\setminus\mathbb{N}$ result this last step also requires MA.

In this chapter we will focus on the final local-to-global step. For any unitary $U \in \mathcal{C}(l^2)$ let $\phi_U : \mathcal{C}(l^2) \to \mathcal{C}(l^2)$ be the associated automorphism defined by $\phi_U(X) = U^*XU$. Let a *coherent family of unitaries* be a choice of a unitary $U_E \in \mathcal{C}(l^2)$, for each block sequence E, with the property that ϕ_{U_E} and $\phi_{U_{\tilde{E}}}$ agree on $\pi(\mathcal{D}_E^+) \cap \pi(\mathcal{D}_{\tilde{E}}^+)$, for any two block sequences E and $\tilde{E}$. (Recall that π is the natural projection from $\mathcal{B}(l^2)$ onto $\mathcal{C}(l^2)$. It was effectively this kind of structure that we used in Theorem 28.4 to produce an outer automorphism.) We will explain how OCA turns any coherent family of unitaries into a single unitary.

We begin by abelianizing the problem. Let $E_0 = (0, 1, 2, \ldots)$, so that $\mathcal{D}_0 = \mathcal{D}_{E_0}$ is the standard diagonal subalgebra of $\mathcal{B}(l^2)$.

Lemma 29.1. *Let (U_E) be a coherent family of unitaries. Then $(U_{E_0}^* U_E)$ is a coherent family of unitaries, each of the form $\pi(M_{\vec{u}})$ for some $\vec{u} \in \mathbb{T}^\mathbb{N}$.*

Proof. The first assertion is straightforward. For the second assertion,

113

observe that coherence implies that each $\phi_{U^*_{E_0}U_E}$ agrees with $\phi_{U^*_{E_0}U_{E_0}} = \mathrm{id}_{\mathcal{C}(l^2)}$ on $\pi(\mathcal{D}_0)$ and hence fixes $\pi(\mathcal{D}_0)$ pointwise. Thus $U^*_{E_0}U_E$ commutes with every element of $\pi(\mathcal{D}_0)$. But the latter is a maximal abelian subalgebra of $\mathcal{C}(l^2)$, so we infer that $U^*_{E_0}U_E \in \pi(\mathcal{D}_0)$. The fact that every unitary in $\pi(\mathcal{D}_0)$ is the image of a unitary in $\mathcal{D}_0$ is standard; for a direct proof, suppose $\pi(\vec{a})$ is unitary in $\mathcal{C}(l^2)$, show that we must have $|a_n| \to 1$, and conclude that $\pi(\vec{a}) = \pi(\vec{u})$ where $u_n = \frac{a_n}{|a_n|}$ (taking $\frac{0}{0} = 1$). $\qquad\square$

Thus, if we know that a given automorphism of $\mathcal{C}(l^2)$ comes from a coherent family of unitaries, we can compose it with $\phi_{U^*_{E_0}}$ to get an automorphism that comes from a coherent family of unitaries of the form $\pi(M_{\vec{u}})$. Since the composition of two inner automorphisms is inner, it will suffice to show that the latter automorphism is inner.

We need one more elementary lemma. Let $\mathbb{N}^{\uparrow\mathbb{N}}$ be the set of all strictly increasing functions $f : \mathbb{N} \to \mathbb{N}$ such that $f(0) > 0$. Recall that $f \leq_* g$ means $f(n) \leq g(n)$ for all but finitely many n; say that $X \subseteq \mathbb{N}^{\uparrow\mathbb{N}}$ is *cofinal* if for every $f \in \mathbb{N}^{\uparrow\mathbb{N}}$ there exists $g \in X$ with $f \leq_* g$. This is the same as saying that X is dense in $\mathbb{N}^{\uparrow\mathbb{N}}$.

Lemma 29.2. *Let X be a cofinal subset of $\mathbb{N}^{\uparrow\mathbb{N}}$.*

(a) If $X = \bigcup X_n$ then some X_n is cofinal.
(b) For infinitely many values of n there exists an i such that there are functions $f \in X$ with $f(i) \leq n$ and $f(i+1)$ arbitrarily large.

Proof. (a) Suppose no X_n is cofinal, and for each n find $f_n \in \mathbb{N}^{\uparrow\mathbb{N}}$ such that $f_n \not\leq_* g$ for any $g \in X_n$. Then diagonalizing the sequence (f_n) yields a function which falsifies cofinality of X.

(b) We first claim that for infinitely many values of n, for every k there exist $i \in \mathbb{N}$ and $f \in X$ such that $f(i) \leq n$ and $f(i+1) \geq k$. Suppose not; then for some n_0, no $n \geq n_0$ satisfies the stated condition. That is, for every $n \geq n_0$ there exists $g(n) = k$ such that every i and f satisfy either $f(i) > n$ or $f(i+1) < g(n)$. For each $m \in \mathbb{N}$ define $h_m : \mathbb{N} \to \mathbb{N}$ by setting $h_m(0) = \max(m, n_0)$ and $h_m(i+1) = g(h_m(i))$. Then for any $f \in X$ a simple induction on i shows that $f(i) \leq h_{f(0)}(i)$ for all i, so that diagonalizing the sequence (h_m) yields a function which contradicts cofinality of X. This proves the claim.

Now the condition $f(i) \leq n$ implies that $i \leq n$. Thus in the claim some value of i must work for infinitely many k, and hence for all k. $\qquad\square$

For any $f \in \mathbb{N}^{\uparrow\mathbb{N}}$ define $\hat{f} \in \mathbb{N}^{\uparrow\mathbb{N}}$ by $\hat{f}(k) = f^k(0)$. (This is why we require $f(0) > 0$.) The key property of $\hat{f}$ is that if $f(k) \geq g(k)$ for all

k then for any i with $g(i) > f(0)$ there exists k such that $\hat{f}(k) < g(i) < g(i+1) \le \hat{f}(k+2)$. For if $\hat{f}(k) < g(i) \le \hat{f}(k+1)$ then

$$\hat{f}(k+2) = f(\hat{f}(k+1)) \ge f(g(i)) \ge f(i+1) \ge g(i+1).$$

We write E_f for the block sequence whose terms are $\hat{f}(k)$.

Theorem 29.3. *Assume OCA. Let ϕ be an automorphism of $\mathcal{C}(l^2)$ and suppose there is a coherent family of unitaries (U_E) such that ϕ agrees with ϕ_{U_E} on $\pi(\mathcal{D}_E^+)$, for every block sequence E. Then ϕ is inner.*

Proof. By Lemma 29.1 we can reduce to the case that for every block sequence E there exists $\vec{u} \in \mathbb{T}^{\mathbb{N}}$ such that ϕ agrees with $\phi_{\pi(M_{\vec{u}})}$ on $\pi(\mathcal{D}_E^+)$. Now give $\mathbb{N}^{\uparrow\mathbb{N}}$ the topology it inherits from $\mathbb{N}^{\mathbb{N}}$ and define $\mathcal{X} \subset \mathbb{N}^{\uparrow\mathbb{N}} \times \mathbb{T}^{\mathbb{N}}$ to be the set of pairs $\langle f, \vec{u}\,\rangle$ such that ϕ agrees with $\phi_{\pi(M_{\vec{u}})}$ on $\pi(\mathcal{D}_{E_f}^+)$. For any $\vec{u}, \vec{v} \in \mathbb{T}^{\mathbb{N}}$ and any $J \subset \mathbb{N}$ define

$$\delta_J(\vec{u}, \vec{v}) = \sup_{i,j \in J} |u_i \bar{u}_j - v_i \bar{v}_j|;$$

then fix $\epsilon > 0$ and let $K_\epsilon \subset \mathcal{X}^2 \setminus \Delta_{\mathcal{X}}$ be the set of pairs $\langle f, \vec{u}\rangle$, $\langle g, \vec{v}\rangle$ for which there exist k and k' such that

$$J = [\hat{f}(k), \hat{f}(k+1)) \cap [\hat{g}(k'), \hat{g}(k'+1))$$

is nonempty and $\delta_J(\vec{u}, \vec{v}) > \epsilon$. This is an open coloring.

We show that $\mathcal{X}$ has no uncountable 0-homogeneous sets for K_ϵ. To see this suppose $Y \subseteq \mathcal{X}$ has cardinality $\aleph_1$. Let $\hat{Y}^1 = \{\hat{g} \mid \langle g, \vec{v}\rangle \in Y$ for some $\vec{v}\}$; then $\hat{Y}^1$ is bounded in $\mathbb{N}^{\mathbb{N}}$ by Theorem 27.2. So there exists $f \in \mathbb{N}^{\uparrow\mathbb{N}}$ with $\hat{g} \le_* f$ for all $\hat{g} \in \hat{Y}^1$. Thus, for each $\hat{g} \in \hat{Y}^1$ there exists N such that $\hat{g}(n) \le f(n)$ for all $n \ge N$. Without loss of generality (possibly replacing Y with an uncountable subset of Y) we can assume the same value of N works for every $\hat{g} \in \hat{Y}^1$, and then by increasing f we can arrange that $\hat{g}(n) \le f(n)$ for all $\hat{g} \in \hat{Y}^1$ and all $n \in \mathbb{N}$. Choose $\vec{u} \in \mathbb{T}^{\mathbb{N}}$ such that ϕ agrees with $\phi_{\pi(M_{\vec{u}})}$ on $\pi(\mathcal{D}_{E_f}^+)$. Now for any $\langle g, \vec{v}\rangle \in Y$ the comment we made just before the theorem shows that any interval $[\hat{g}(i), \hat{g}(i+1))$ is contained in an interval $[\hat{f}(k), \hat{f}(k+2))$, so the fact that ϕ agrees with both $\phi_{\pi(M_{\vec{u}})}$ and $\phi_{\pi(M_{\vec{v}})}$ on $\pi(\mathcal{D}_{E_g}^+)$ implies that $\delta_{[\hat{g}(i),\hat{g}(i+1))}(\vec{u}, \vec{v}) \to 0$ as $i \to \infty$; if the limit were not zero then we could construct an operator $B \in \mathcal{D}_{E_g}^+$ as in Lemma 28.3 such that $\phi_{\pi(M_{\vec{u}})}$ and $\phi_{\pi(M_{\vec{v}})}$ disagree on $\pi(B)$.

For some $k, m \in \mathbb{N}$ there must be uncountably many $\langle g, \vec{v}\rangle \in Y$ such that $\hat{g}(m+1) = k$ and $\delta_{[\hat{g}(i),\hat{g}(i+1))}(\vec{u}, \vec{v}) \le \epsilon/2$ for $i \ge m$. Since $\mathbb{T}^m$ is separable we can find distinct $\langle g, \vec{v}\rangle, \langle g', \vec{v}'\rangle \in Y$ such that $\hat{g}(i) = \hat{g}'(i)$ for

$i \leq m + 1$ and $|u_i - v_i| \leq \epsilon/2$ for $i \leq k$. It follows that $\delta_J(\vec{u}, \vec{v}) \leq \epsilon$ whenever $J = [\hat{f}(k), \hat{f}(k+1)) \cap [\hat{g}(k'), \hat{g}(k'+1))$ is nonempty. Thus Y is not 0-homogeneous for K_ϵ, so we conclude that there are no uncountable 0-homogeneous sets for K_ϵ.

OCA now implies that for every $\epsilon > 0$ the space $\mathcal{X}$ is covered by countably many sets which are 1-homogeneous for K_ϵ. Use Lemma 29.2 (a) to recursively find a sequence $X_0 \supseteq X_1 \supseteq X_2 \supseteq \cdots$ such that each X_n is 1-homogeneous for $K_{2^{-n}}$ and each $\hat{X}_n^1 = \{\hat{f} \mid \langle f, \vec{u} \rangle \in X_n \text{ for some } \vec{u}\}$ is cofinal in $\mathbb{N}^{\uparrow\mathbb{N}}$. By Lemma 29.2 (b) we can find an increasing sequence (m_n) such that for each n there exists a value of i and functions $\hat{f} \in \hat{X}_n^1$ with $\hat{f}(i) \leq m_n$ and $\hat{f}(i+1)$ arbitrarily large.

For each n and j in $\mathbb{N}$ find $\langle f_{n,j}, \vec{u}^{n,j} \rangle \in X_n$ and $i \in \mathbb{N}$ such that $\hat{f}_{n,j}(i) < m_n < m_{n+j} \leq \hat{f}_{n,j}(i+1)$. Then for each n extract a subsequence so that $\vec{u}^{n,j}$ converges in $\mathbb{T}^{\mathbb{N}}$ and let $\vec{u}^n$ be the limit. We claim that $n < n'$ implies $\delta_{[m_{n'},\infty)}(\vec{u}^n, \vec{u}^{n'}) \leq 2^{-n}$. To see this, suppose $n_1, n_2 \geq m_{n'}$; then combine the facts that (1) $\hat{f}_{n,j}, \hat{f}_{n',j} \in \hat{X}_n^1$ for all j; (2) X_n is 1-homogeneous for $K_{2^{-n}}$; (3) for sufficiently large j there exist i and i' with $\hat{f}_{n,j}(i) < n_1, n_2 < \hat{f}_{n,j}(i+1)$ and $\hat{f}_{n,j'}(i') < n_1, n_2 < \hat{f}_{n,j'}(i'+1)$; and (4) $\vec{u}^{n,j}$ and $\vec{u}^{n',j}$ converge pointwise to $\vec{u}^n$ and $\vec{u}^{n'}$. This yields the claim.

Now we have to align the sequence $\vec{u}^n = (u_k^n)$. For all $n < n'$ define $z_{n,n'} = u_{m_{n'}}^{n'} \bar{u}_{m_{n'}}^n$. The previous claim then yields that $|z_{n,n'} u_k^n - u_k^{n'}| \leq 2^{-n}$ for all $k \geq m_{n'}$. If $n < n' < n''$, then taking $k = m_{n''}$ shows that $|z_{n,n'} z_{n',n''} - z_{n,n''}| \leq 3 \cdot 2^{-n}$. An application of infinite Ramsey theory then yields a strictly increasing sequence (n_k) such that $|z_{n_k, n_{k'}} - 1| \leq 4 \cdot 2^{-k}$ for all $k < k'$. Now define $\vec{w} \in \mathbb{T}^n$ by letting it agree with $\vec{u}^{n_k}$ on the interval $[m_{n_k}, m_{n_{k+1}})$. Then for any $\langle f, \vec{u} \rangle \in X_{n_k}$, by 1-homogeneity and the definition of $\vec{u}^{n_k}$ we have that $\vec{u}$ and $\vec{u}^{n_k}$ approximately agree on $[m_{n_k}, \infty)$ in the sense that $\delta_J(\vec{u}, \vec{u}^{n_k}) \leq 2^{-k}$ whenever $J = [\hat{f}(i), \hat{f}(i+1))$ satisfies $\hat{f}(i) \geq m_{n_k}$. A short computation shows that $\vec{u}$ and $\vec{w}$ also approximately agree in the same sense with a slightly worse constant, namely $5 \cdot 2^{-k}$. This means that $\|\phi_{\pi(M_{\vec{u}})}(B) - \phi_{\pi(M_{\vec{w}})}(B)\| \leq 5 \cdot 2^{-k}\|B\|$ for all $B \in \pi(\mathcal{D}_{E_f})$. Since k was arbitrary and X_{n_k} is cofinal in $\mathbb{N}^{\uparrow\mathbb{N}}$, Lemma 28.1 implies that $\phi(B) = \pi(M_{\vec{w}})^* B \pi(M_{\vec{w}})$ for all $B \in \mathcal{C}(l^2)$. Thus ϕ is inner. $\square$

Chapter 30

The Multiverse Interpretation

What are we to make of the independence phenomena evinced by forcing?

The two kinds of independence mentioned in Chapter 1 in geometry and number theory offer us strikingly different paradigms. In both cases there is broad agreement about the correct interpretation of the independence results. For instance, no one nowadays would consider it meaningful to ask whether the parallel postulate is "really true" in some universal sense; it simply holds in some two-dimensional geometries and fails in others.

In contrast, although the arithmetical expression of the consistency of PA is independent of PA, it is still widely regarded as true. To take a more extreme example, consider the formal system MC = ZFC + "measurable cardinals exist". Few would suggest that the sentence Con(MC) which arithmetically expresses that MC is a consistent system might lack a well-defined truth value. Yet Con(MC) is presumably independent of PA, indeed, presumably even independent of ZFC.

The reason for this discrepancy is probably because we see the concept *point in space* as depending on a choice of ambient space, whereas the concept *natural number* has no comparable ambiguity. PA may have nonstandard models in which the truth value of Con(MC) varies, but it has only one standard model, and Con(MC) has only one truth value there.

We have an almost physical intuition for elementary number theory. One can imagine mechanically evaluating whether MC is consistent by simply running through all possible formal derivations in MC and checking for an inconsistency. Of course this involves an infinitely long computation, but that need not place it outside the realm of conceivable possibility. Davies has described an appealing scenario in which a machine is programmed to respond to being input the natural number n by first performing the nth step of some computation and then building a duplicate copy of itself which is half as large and runs twice as fast, and feeding it the input $n + 1$. The envisioned result is a cascade of ever-smaller and -faster machines which collectively occupy a finite volume and perform an infinite computation in a finite amount of time. For instance, they could check for an inconsistency in MC. Any discovery of an inconsistency could be relayed back up the chain, so that the initial machine could report a definite result after the entire computation was complete. Although one doubts such a scenario is compatible with the physics of our universe, it still seems cogent. Presumably number theory skeptics would disagree, but is the idea of a Davies type computation really any less comprehensible than, say, a computation of length 10^{100} (which might very well also be incompatible with the physics of our universe)?

If the concept of a countably infinite computation is legitimate, then the truth value of Con(MC), or indeed of any sentence in the language of PA, could in principle be mechanically evaluated. This would seem to compel us to regard these truth values as being well-defined, regardless of whether they could be settled within PA, or any other particular formal system.

All this is a preface to the question of whether set theory better resembles number theory or geometry with regard to independence phenomena. Should we suppose that the continuum hypothesis, for example, has a definite truth value in a well-defined canonical model? Or is there a range of models in which the truth value of the continuum hypothesis varies, none of which has any special ontological priority?

Forcing tends to push us in the latter direction. It creates the impression that there is a range of equally valid models of ZFC, and that one can always pass to a larger model in which the value of $2^{\aleph_0}$ changes. Now according to ZFC^+, this variety only attaches to countable models, and when CH is evaluated in the class of all sets it still has a well-defined truth value. On the other hand, ZFC^+ itself presumably has countable models to which one can apply the forcing construction, so perhaps it ought not to be trusted on this point. In an influential series of recent papers, Hamkins has vigorously argued for the position that there is no canonical model of ZFC, a position that he calls "the multiverse view". A version of this multiverse view goes back to Skolem, who called it a "relativity of set theoretic notions".

One potential difficulty for the multiverse interpretation comes from Davies's infinite machines. If we could form a clear conception of some infinite computation that would settle the continuum hypothesis, for example, this would suggest that we ought to conclude it has a well-defined truth value. What about evaluating CH by systematically examining every function from $\aleph_1$ to $\mathcal{P}(\mathbb{N})$ to see if one of them is surjective? The problem with this idea is that it is not obviously cogent if we have not yet adopted ZFC as a universal background theory. How would one go about systematically examining every function from $\aleph_1$ to $\mathcal{P}(\mathbb{N})$? There is no plausible Davies machines type story to help us get there in this case. So using this kind of idea as a justification for adopting ZFC as a universal background theory is problematic.

A more general philosophical objection to the multiverse interpretation is that it presumes some ambiguity in our set concept. According to this argument, the multiverse picture is ill-conceived because, just as in the case of PA, there is a well-defined canonical model of ZFC. ZFC models sets, a set is nothing more than a collection, and *collection* is a purely logical concept that is completely unambiguous.

In light of the paradoxes of naive set theory, this last statement is surely too bold. At the very least, we have to admit that there are two versions of the informal notion of a collection, namely, sets and classes. Moreover, the exact nature of the conceptual distinction between these two versions has been elusive. At some crude level we recognize that proper classes are "too big" to be sets, but this is hardly an adequate explanation. If we cannot give a convincing account of how the informal notion of a collection comes to have two distinct mathematical expressions, then we might not be in a good position to criticize the multiverse theorist's claim that there are a multitude of distinct set concepts.

I think there actually is a reasonable explanation of the difference between sets and classes. When we first learn about sets we are told that there are two ways of specifying a set, either by explicitly exhibiting its elements or by stating a rule which determines whether any given individual does or does not belong to the set. It might be thought that this distinguishes finite sets, which can be explicitly exhibited, from infinite sets, which cannot. However, the theoretical posibility of Davies machines calls this assumption into question. If infinite computations are possible, in principle, then it should also be possible, in principle, to exhibit (to survey, to perform a computation that runs through) all the elements of at least some infinite sets.

In that case, might every collection be "surveyable" in some theoretical sense? No. However one might concretely represent sets (perhaps by representing a set as a tree, and then encoding the order relation on the tree via an infinite array of 0's and 1's), there is no conceivable scenario in which the collection of all sets could be surveyed in its entirety. That is because any collection of sets that could be surveyed could then be amalgamated into a new set that did not appear in the survey. This suggests that the right way to conceptually differentiate between sets and proper classes is to say that sets are collections which can, in principle, be surveyed, whereas proper classes are collections that can only be specified indirectly. But if this thesis is granted, then the objection to the multiverse view from conceptual definiteness runs into the same difficulty as the objection from infinite machines. Anyone who has not already accepted ZFC as a universal background theory is bound to find it difficult to imagine surveying, say, all sets of natural numbers. The point may be contentious because the converse is also true: someone who has already accepted ZFC is likely to consider the idea of surveying $\mathcal{P}(\mathbb{N})$ unproblematic. But that begs the question. If we first get the surveyable/unsurveyable distinction straight, and then consider how to axiomatize set theory, then the natural reaction to ZFC is that it does not capture the concept of a surveyable collection because the power set axiom converts surveyable collections into unsurveyable ones.

A picture emerges of a mathematical universe which is composed of countable structures that have absolute properties and which includes a range of countable models of ZFC in which the truth values of questions like the continuum hypothesis can vary. Thus, with regard to independence phenomena, if we take "set theory" to be the theory of surveyable collections then it has an absolute meaning and behaves like number theory, but questions like the continuum hypothesis cannot even be posed; if we take it to be the theory of individuals in some model of ZFC then it has a variable meaning and behaves like geometry.

This conclusion is corroborated by our experience with ordinary mathematics. We certainly need uncountable collections such as the real line in order to do normal mathematics, but we only need them as classes, not as sets. Situations in which one genuinely requires uncountably long constructions are rare. To the contrary, it is well established that the vast bulk of mainstream mathematics is absolute and can be formalized in essentially number-theoretic systems. This is true even in fields like functional analysis that have a set-theoretic flavor, provided one sticks to the separable/separable predual case of primary interest.

The reader will have noticed how peripheral the independence results discussed in this book are to mainstream mathematical concerns. Normal mathematics is absolute.

Appendix A

Forcing with Preorders

In this book we have defined forcing notions concretely as families of sets ordered by inclusion. The standard definition treats them abstractly as preordered sets. Our version simplifies the exposition a little because the order relation is built in and does not have to be added separately. It may appear to be less general than the preorder definition, but the two are actually equivalent. The purpose of this appendix is to explain this equivalence.

Recall that a *preordered set* is a set equipped with a relation that is reflexive and transitive, but not necessarily antisymmetric — we can have $p \leq q$ and $q \leq p$ for distinct p and q. (Beware: in the forcing literature the term "partial order" sometimes means "preorder".) In the abstract approach, it is convenient to use preorders when we get to iterated forcing. The notions of *extension, compatible elements, dense set, ideal,* and *generic ideal* all generalize straightforwardly to preordered sets, just as in Chapter 23.

We should also mention that according to the usual convention an extension of p lies below p, not above it, and one is correspondingly interested in the dual notion of filters rather than ideals. We have reversed this convention in order to avoid habitually ordering sets by reverse inclusion. Having forcing notions grow downward carries no particular benefit, although it does make sense in the context of Boolean-valued forcing.

Preorders do not create any genuinely greater generality over partial orders. We can always factor out the equivalence relation which sets $p \sim q$ if $p \leq q$ and $q \leq p$, and it is more or less obvious that forcing with the resulting poset is equivalent in every significant sense to forcing with the original preordered set. It should also be clear that replacing one poset with another one to which it is order-isomorphic does not meaningfully affect the forcing construction.

But every nonempty poset $\mathcal{P}$ is order-isomorphic to a forcing notion, according to our definition, ordered by inclusion. Just define $p^\le = \{q \in \mathcal{P} \mid q \le p\}$ for all $p \in \mathcal{P}$ and let $P = \{p^\le \mid p \in \mathcal{P}\}$. The order-isomorphism between $\mathcal{P}$ and P is given by the correspondence $p \leftrightarrow p^\le$.

It should now be clear that forcing with forcing notions as we have defined them entails no loss of generality.

Exercises

Chapter 1

1. Express the assertion that the set of primes contains arithmetical progressions of any finite length in the language of PA.

2. Informally deduce from the non-logical axioms of PA that $x + y = y + x$ for all natural numbers x and y.

3. Informally deduce from the non-logical axioms of PA that $x \cdot (y + z) = x \cdot y + x \cdot z$ for all natural numbers x, y, and z.

Chapter 2

1. Express a few of the Zermelo-Fraenkel axioms in the formal language of set theory.

2. Show that the replacement scheme implies the separation scheme.

3. Give an informal argument that if ZFC is inconsistent then there is a proof in PA that ZFC is inconsistent.

Chapter 3

1. Show that every countable well-ordered set is order-isomorphic to a subset of $\mathbb{Q}$.

2. For any well-ordered set W, let W^ω be the set of all functions $f : \mathbb{N} \to W$ with the property that $f(n)$ is the least element of W for all but finitely many n, ordered by setting $f < g$ if $f(n) < g(n)$ where n is the greatest point at which they differ. Show that W^ω is well-ordered.

3. In the notation of the previous exercise, show that there is a countable well-ordered set W such that W and W^ω are order-isomorphic.

Chapter 4

1. Use Theorem Scheme 4.5 to prove that the transitive closure of any set exists.

2. Prove that $x \in V_{<\omega}$ if and only if $TC(x)$ is finite (in which case we say that x is *hereditarily finite*).

3. Prove that the rank of any ordinal is itself.

Chapter 5

1. (Schroeder-Bernstein theorem) Suppose there are injective maps from X into Y and from Y into X. Prove that there is a bijection between X and Y.

2. Prove a few cases of Proposition 5.5.

3. (König's theorem) Let (X_n) and (Y_n) be two infinite sequences of sets such that $\text{card}(X_n) < \text{card}(Y_n)$ for all n. Prove that $\text{card}(\bigcup X_n) < \text{card}(\prod Y_n)$. Infer that $2^{\aleph_0} \neq \aleph_\omega$.

Chapter 6

1. A set x is *hereditarily countable* if $TC(x)$ is countable. Prove a theorem scheme which states that the set of hereditarily countable sets satisfies all the axioms of ZFC except the power set axiom.

2. Define *restricted quantifiers* by setting $(\forall x \in y)\phi \equiv (\forall x)(x \in y \to \phi)$ and $(\exists x \in y)\phi \equiv (\exists x)(x \in y \wedge \phi)$ (where any variables can take the place of x and y). Verify that $(\exists x \in y)\phi$ is equivalent to $\neg(\forall x \in y)\neg\phi$.

3. A Δ_0 *formula* is a formula which only involves restricted quantifiers. Prove that every Δ_0 formula in the language of set theory is absolute for any transitive set. Express the statements "x is an ordinal" and "x is a subset of $\mathbb{N}$" as Δ_0 formulas.

Chapter 7

1. The language of group theory is defined analogously to the language of Peano arithmetic, except that terms are built up using a constant symbol e and symbols for the product and inverse operations. Let ϕ be a sentence in the language of group theory and suppose there is a group G which satisfies ϕ. Prove that G contains a countable subgroup that also satisfies ϕ.

2. Devise a formal system ZFC^{++} in which we can reason about a countable transitive set $\mathbf{M}'$ which satisfies all the axioms of ZFC^+.

3. Prove a metatheorem which says that if ZFC is consistent then so is ZFC^{++}, as in the previous exercise.

Chapter 8

1. Let $X, Y \in \mathbf{M}$ and let P be the set of all finite partial functions from X to Y. Prove that for any generic ideal G of P there is a function $\tilde{f} : X \to Y$ such that the elements of G are precisely the restrictions of $\tilde{f}$ to finite subsets of X.

2. Say that a forcing notion P is *complete* if whenever $x \in P$ and $y \in \mathbf{M}$ is a subset of x, we have $y \in P$. Prove that if P is complete and $x, y \in P$ are compatible then $x \cup y \in P$. Define the *completion* of any forcing notion P to be the set $\overline{P} = \{ y \in \mathbf{M} \mid y \subset x \text{ for some } x \in P \}$. Prove that the map $G \mapsto G \cap P$ takes the generic ideals of $\overline{P}$ bijectively onto the generic ideals of $\mathcal{P}$.

3. Let P be the set of all finite partial functions from $\mathbb{N}^2$ into $\{0, 1\}$, let G be a generic ideal of P, let $\tilde{f} : \mathbb{N}^2 \to \{0, 1\}$ be the corresponding function, and let $R = \tilde{f}^{-1}(1) \subset \mathbb{N}^2$ be the generic relation on $\mathbb{N}$ of which $\tilde{f}$ is the characteristic function. Prove that every finite partially ordered set is isomorphic to some finite subset of $\mathbb{N}$ equipped with the restriction of R. Prove there is an infinite sequence (n_k) such that $(n_j, n_k) \in R$ if and only if $j < k$.

Chapter 9

1. Let P be a forcing notion, let G be a generic ideal of P, and suppose $G \in \mathbf{M}$. (As we noted at the beginning of the chapter, this can only happen if P is trivial.) Prove that $\mathbf{M}[G] = \mathbf{M}$.

2. Let P be the set of all finite partial functions from $\mathbb{N}^2$ into $\{0, 1\}$, let α be a countable ordinal that does not belong to $\mathbf{M}$, and let $R \subset \mathbb{N}^2$ be a well-ordering of $\mathbb{N}$ that makes it order-isomorphic to α. Let G be the set of restrictions of the characteristic function of R to finite subsets of $\mathbb{N}^2$, as in Exercise 8.3. Show that $\mathbf{M}[G]$ does not model the axioms of ZFC. Specifically, which axiom fails?

3. Let P be the forcing notion from Exercise 8.1 and let $\tilde{f}$ be the function associated to a generic ideal. Exhibit a P-name whose value is the function $\tilde{f}$.

Chapter 10

1. Let $x \in \mathbf{M}$ and let τ be a P-name every element of which has the form $\langle \check{y}, p \rangle$ for some $y \in \mathbf{M}$ and $p \in P$. Let G be a generic ideal of P. Prove that $x \in \tau^G$ if and only if $p \Vdash \check{x} \in \tau$ for some $p \in G$.

2. Let τ_1 and τ_2 be P-names. Suppose every element of τ_1 has the form $\langle \check{x}, p \rangle$ for some $x \in \mathbf{M}$ and $p \in P$, and similarly for τ_2. Characterize which $p \in P$ force $\tau_1 \subseteq \tau_2$.

3. With the assumptions in the previous exercise, prove that $\tau_1^G \subseteq \tau_2^G$ if and only if $p \Vdash \tau_1 \subseteq \tau_2$ for some $p \in G$.

Chapter 11

1. Prove the two claims in the proof of Theorem Scheme 11.1, for each possible form of ϕ.

2. Exhibit a P-name τ such that $\tau^G = P \setminus G$ for any generic ideal G.

3. Let G be a generic ideal, let $x \in \mathbf{M}$, and suppose $y \subseteq x$ lies in $\mathbf{M}[G]$. Show that there is a P-name τ all of whose elements are of the form $\langle \check{z}, p \rangle$ with $z \in x$ and $p \in P$ and such that $\tau^G = y$.

Chapter 12

1. Let $X, Y \in \mathbf{M}$ be infinite and let P be the set of finite partial bijections between X and Y. Show that if $A \in \mathbf{M}$ is infinite then $\mathcal{P}(A)^{\mathbf{M}[G]} \neq \mathcal{P}(A)^{\mathbf{M}}$ for any generic ideal G of P.

2. A subset S of a forcing notion P is *finitely compatible* if any finite set of elements in S have a common extension. Suppose $\mathbf{M} \models$ "the union of every finitely compatible subset of P of cardinality at most $\aleph_1$ belongs to P". Prove that for any $X \in \mathbf{M}$ and any generic ideal G of P, every function $f : \aleph_1 \to X$ in $\mathbf{M}[G]$ is already in $\mathbf{M}$.

3. Prove that it is relatively consistent with ZFC that $2^{\aleph_1} = \aleph_2$.

Chapter 13

1. Prove that every countable forcing notion preserves cardinals.

2. Given a graph with $\aleph_2$ vertices and such that each vertex has at most $\aleph_1$ neighbors, show that there is a set of $\aleph_2$ vertices no two of which are neighbors.

3. Let $X \in \mathbf{M}$ be infinite, let P be the set of all finite partial functions from X into $\{0, 1\}$, and let G be a generic ideal of P. Prove that P preserves cardinals and $\mathbf{M}[G] \models 2^{\aleph_0} \geq \mathrm{card}(X)$.

Chapter 17

1. Prove that the intersection of any countable family of club subsets of $\aleph_1$ is a club subset of $\aleph_1$. In particular, every club set is stationary.

2. Prove that $\Diamond$ implies CH.

3. Suppose we are given a sequence $\{ S_\alpha \mid \alpha < \aleph_1 \}$ of countable subsets of $\aleph_1$. Prove that if $\{ h_\alpha \mid \alpha < \aleph_1 \}$ is any complete vertex sequence, we can find a path f up the standard $\aleph_1$-$\aleph_1$ tree such that $f|_\alpha \neq h_\alpha$ for any α satisfying $\alpha \in \bigcup_{\beta < \alpha} S_\beta$.

Chapter 20

1. Prove a converse to Theorem 20.4 which states that if G_1 is a generic ideal of P_1 relative to $\mathbf{M}$ and G_2 is a generic ideal of P_2 relative to $\mathbf{M}[G_1]$, then $G_1 \cdot G_2$ is a generic ideal of $P_1 \cdot P_2$ relative to $\mathbf{M}$.

2. Suppose $\mathbf{M} \models 2^{\aleph_1} = \aleph_2$. Use Lemma 20.6 to prove that the model constructed in Theorem 13.5 satisfies $2^{\aleph_0} = \aleph_2$.

3. Generalize Lemma 20.5 to higher cardinalities.

Chapter 22

1. How should the definitions of G_1 and G_2 in Theorem 22.3 be changed if P and π are not assumed to be rooted?

2. Let P be a rooted forcing notion for $\mathbf{M}$, let π be a rooted P-name, let G_1 be an ideal of P which is generic relative to $\mathbf{M}$, and let G_2 be an ideal of π^{G_1} which is generic relative to $\mathbf{M}[G_1]$. Prove that $G = G_1 * G_2$, as defined in the proof of Theorem 22.3, is an ideal of $P * \pi$ which is generic relative to $\mathbf{M}$.

3. In the notation of Theorem 22.5, show that $G_{\beta+1} = G_\beta * G'_\beta$ (as defined in the proof of Theorem 22.3) for all β.

Chapter 23

1. Let P be an ω-closed forcing notion and let $\{D_\alpha\}$ be a family of $\aleph_1$ many dense subsets of P. Prove that there is an ideal of P that intersects every D_α.

2. Let F be a family of nonempty subsets of $\mathbb{N}$ any two of which have finite intersection. Suppose $\operatorname{card}(F) < 2^{\aleph_0}$. Assuming MA, prove that F is not maximal.

3. A cardinal κ is *regular* if it is not the union of fewer than κ many sets each of cardinality less than κ. Suppose we are given a κ-stage finite support iteration construction such that $\mathbf{M} \models$ "P^* is c.c.c. and κ is regular." Let G be a generic ideal of P^*, let $A \in \mathbf{M}$, let $B \in \mathbf{M}[G]$ be a subset of A, and assume $\mathbf{M} \models \operatorname{card}(A) = \kappa$. In the notation of Theorem 22.5, show that B belongs to $\mathbf{M}[G_\beta]$ for some $\beta < \kappa$.

Chapter 26

1. For any separable metric space $\mathcal{X}$, prove that there is a continuous bijection between a subset of the Cantor set and $\mathcal{X}$. Infer that the OCA for subsets of $\mathbb{R}$ implies the OCA for all separable metric spaces.

2. (Infinite Ramsey theorem) Let X be an infinite set and let $K \subseteq X^2 \backslash \Delta_X$. Prove that there is an infinite subset $Y \subseteq X$ such that either $\langle x, y \rangle \in K$ for all distinct $x, y \in Y$ or $\langle x, y \rangle \notin K$ for all distinct $x, y \in Y$.

3. Show that there exists $K \subset \mathbb{R}^2 \setminus \Delta_{\mathbb{R}}$ with respect to which there are no uncountable 0-homogeneous or 1-homogeneous sets.

Notes

Chapter 1: Euclid's *Elements* are available in Dover paperback [11]. There are a number of excellent textbooks on non-euclidean geometry. Peano first presented his axioms in [34], although essentially the same system was proposed earler by Dedekind (see [8]). Peano arithmetic is discussed in every good introduction to mathematical logic, for instance in Mendelson [29]. Gödel's independence results can be found there too. For simple arithmetical statements that are independent of PA see [24].

Chapter 2: Zermelo's original axioms appear in [51]. Fraenkel's addition [14] was independently proposed by Skolem [40]. Two standard treatments of axiomatic set theory are the books by Jech [21] and Kunen [26]. However, for more on how to use PA to formalize reasoning about provability in ZFC one should go back to Mendelson or something similar.

Chapter 3: Zermelo's proof of the well-ordering theorem originally appeared in [50].

Chapter 4: The definition of ordinals is due to von Neumann [48]. It is stated in one of his earliest published papers, allegedly written while he was in high school.

At limit ordinals V_α is usually defined to equal $V_{<\alpha}$ rather than its power set. This goes against the spirit of von Neumann's definition of ordinals, with its uniform treatment of successors and limits, and it leads to a number of minor complications, for instance in the proof of Proposition 4.6 and in the definition of rank.

Chapter 5: Cantor's basic theory of infinite cardinalities can be found in the monograph [4].

Chapter 6: Mostowski's collapsing lemma appears in [31].

Chapter 7: The reflection principle is due to Montague [30].

Chapter 8: The next several chapters present, in modern form, Cohen's theory of forcing. It was originally expounded at length in [6]. The relation between our definition of a forcing notion and the usual definition is discussed in the appendix.

Chapter 9: Condition (ii) in Definition 9.1 is nonstandard, although, as we explain in the text, it is harmless. This minor change pays off handsomely by simplifying virtually every subsequent forcing argument.

Chapter 12: Gödel's proof of the relative constency of the continuum hypothesis (in fact, the generalized continuum hypothesis) is given in [18].

Chapter 13: See [5] for the original proof of the relative consistency of the negation of the continuum hypothesis.

Chapter 14: Erdös's theorem is proven in [10].

Chapter 15: Stone duality is discussed in [17]. Rudin's theorem can be found in [45].

Chapter 16: The fact that the state $A \mapsto \lim_{n \to \xi} \langle Ae_n, e_n \rangle$ is pure is due to Anderson [3]. Theorem 16.4 is proven in [2].

The pure state $A \mapsto \lim_{n \to \xi} \langle Ae_n, e_n \rangle$ on $\mathcal{B}(H)$ restricts to a pure state on the *diagonal subalgebra* (relative to the basis (e_n)) consisting of the operators whose off-diagonal entries $\langle Ae_m, e_n \rangle$, $m \neq n$, are all zero. According to the spectacular recent solution to the Kadison-Singer problem found by Marcus, Spielman, and Srivastava [27], there is no other state on $\mathcal{B}(H)$ which takes the same values on the diagonal subalgebra.

Chapter 17: The diamond principle is due to Jensen [22]. It is usually stated in terms of the standard tree of height $\aleph_1$ in which each vertex has exactly two immediate successors. In this tree the vertices at level α can be identified with subsets of α. The equivalence of this version to ours is an easy exercise — just restrict attention to limit levels, and use CH to get that each vertex has $\aleph_1$ extensions at the next limit level.

Chapter 18: Jensen's proof that diamond implies the existence of Suslin lines is in [22]. The relative consistency of the existence of Suslin lines had been proven earlier by Jech [20] and Tennenbaum [43].

Chapter 19: Naimark's two papers on the subject are [32] and [33]. Rosenberg [36] showed that there could be no separable counterexamples to Naimark's problem. Theorem 19.4 is from [1]. The homogeneity of the pure state space of a simple separable C*-algebra is proven in [25].

Chapter 20: The stronger diamond principle we present here is also due to Jensen [22]. The idea of using product forcing to prove Theorem 20.7 was suggested to me by Ilijas Farah.

Chapter 21: Theorem 21.1 is due to Stein [42]. Shelah's paper is [37]; another good source is [9]. The relative consistency of a positive answer to

Whitehead's problem for groups of any cardinality is proven in [38].

Chapter 22: The technique of iterated forcing is due to Solovay and Tennenbaum [41].

Chapter 23: Martin's axiom first appeared in [28]. Theorem 23.3 is from [41].

Chapter 24: The standard source for consequences of Martin's axiom is [15]. The version of Theorem 16.4 in which CH is weakened to MA was proven by Farah and the author ([13], Theorem 6.46).

Chapter 25: This material can also be found in [37] and [9].

Chapter 26: The open coloring axiom was proven consistent under a large cardinal assumption in [44]. Its relative consistency without large cardinals was shown in [46]; more details are given in [16].

Chapter 27: Theorems 27.1 and 27.2 are from [44]; the remaining material is from [47]. The relative consistency of all automorphisms of $\beta \mathbb{N} \setminus \mathbb{N}$ being trivial was first proven by Shelah using a large cardinal assumption [39].

Chapter 28: See [49] for more on the intuition of $\mathcal{C}(l^2)$ as a noncommutative analog of $\beta \mathbb{N} \setminus \mathbb{N}$. Theorem 28.4 is from [35]. The argument we give here is based on a proof due to Farah [12].

Chapter 29: Farah's theorem is proven in [12]. The fact used in Lemma 29.1 that $\pi(\mathcal{D}_0)$ is maximal abelian in $\mathcal{C}(l^2)$ is proven in [23].

Chapter 30: Davies's paper is [7]. Hamkins presents his views in, for instance, [19]. For Skolem's argument see [40].

Bibliography

[1] C. Akemann and N. Weaver, Consistency of a counterexample to Naimark's problem, *Proc. Nat. Acad. Sci. USA* **101** (2004), 7522-7525.

[2] ———, $B(H)$ has a pure state that is not multiplicative on any MASA, *Proc. Nat. Acad. Sci. USA* **105** (2008), 5313-5314.

[3] J. Anderson, Extreme points in sets of positive linear maps on $B(H)$, *J. Funct. Anal.* **31** (1979), 195-217.

[4] G. Cantor, *Contributions to the Founding of the Theory of Transfinite Numbers* (Dover, 1952).

[5] P. Cohen, The independence of the continuum hypothesis, I, II, *Proc. Nat. Acad. Sci. USA* **50** (1963), 1143-1148; **51** (1964), 105-110.

[6] ———, *Set Theory and the Continuum Hypothesis* (Benjamin, 1966).

[7] E. B. Davies, Building infinite machines, *British J. Philos. Sci.* **52** (2001), 671-682.

[8] R. Dedekind, *Essays on the Theory of Numbers* (Dover, 1963).

[9] P. C. Eklof, Whitehead's problem is undecidable, *Amer. Math. Monthly* **83** (1976), 775-788.

[10] P. Erdös, An interpolation problem associated with the continuum hypothesis, *Michigan Math. J.* **11** (1964), 9-10.

[11] Euclid, *The Thirteen Books of the Elements*, Vol. 1 (Dover, 1956).

[12] I. Farah, All automorphisms of the Calkin algebra are inner, *Ann. of Math.* **173** (2011), 619-661.

[13] I. Farah and E. Wofsey, Set theory and operator algebras, in *Appalachian Set Theory: 2006-2012*, J. Cummings and E. Schimmerling, eds., London Math. Soc. Lecture Note Series **406** (Cambridge, 2012), 63-120.

[14] A. A. Fraenkel, Zu den Grundlagen der Cantor-Zermeloschen Mengenlehre, *Math. Ann.* **86** (1922), 230-237 (German).

[15] D. H. Fremlin, *Consequences of Martin's Axiom* (Cambridge, 1984).

[16] S. Fuchino, Open coloring axiom and forcing axioms, manuscript (2000).

[17] S. Givant and P. Halmos, *Introduction to Boolean Algebras* (Springer, 2009).

[18] K. Gödel, Consistency-proof for the generalized continuum hypothesis, *Proc. Nat. Acad. Sci. USA* **25** (1939), 220-224.

[19] J. D. Hamkins, The set-theoretic multiverse, *Rev. Symb. Log.* **5** (2012), 416-

449.

[20] T. Jech, Nonprovability of Suslin's hypothesis, *Comment. Math. Univ. Carolinae* **8** (1967), 291-305.

[21] ———, *Set Theory* (Academic Press, 1978).

[22] R. Jensen, The fine structure of the constructible hierarchy, *Ann. Math. Logic* **4** (1972), 229-308.

[23] B. E. Johnson and S. K. Parrott, Operators commuting with a von Neumann algebra modulo the set of compact operators, *J. Funct. Anal.* **11** (1972), 39-61.

[24] L. Kirby and J. Paris, Accessible independence results for Peano arithmetic, *Bull. London Math. Soc.* **4** (1982), 285-293.

[25] A. Kishimoto, N. Ozawa, and S. Sakai, Homogeneity of the pure state space of a separable C*-algebra, *Canad. Math. Bull.* **46** (2003), 365-372.

[26] K. Kunen, *Set Theory: An Introduction to Independence Proofs* (Elsevier, 1980).

[27] A. Marcus, D. Spielman, and N. Srivastava, Interlacing families II: mixed characteristic polynomials and the Kadison-Singer problem, arXiv:1306.3969 (2013).

[28] D. A. Martin and R. M. Solovay, Internal Cohen extensions, *Ann. Math. Logic* **2** (1970), 143-178.

[29] E. Mendelson, *Introduction to Mathematical Logic*, fifth edition (Chapman and Hall/CRC, 2009).

[30] R. Montague, Fraenkel's addition to the axioms of Zermelo, in *Essays on the Foundations of Mathematics*, Y. Bar-Hillel et al., eds. (Magnes Press, 1961), 91-114.

[31] A. Mostowski, An undecidable arithmetical statement, *Fund. Math.* **36** (1949), 143-164.

[32] M. Naimark, Rings with involutions, *Uspehi Matem. Nauk* **3** (1948), 52-145 (Russian).

[33] ———, On a problem in the theory of rings with involution, *Uspehi Matem. Nauk* **6** (1951), 160-164 (Russian).

[34] G. Peano, The principles of arithmetic, presented by a new method, in *Selected Works of Giuseppe Peano*, H. C. Kennedy, ed. and trans. (George Allen & Unwin Ltd., 1973), 101-134.

[35] N. C. Phillips and N. Weaver, The Calkin algebra has outer automorphisms, *Duke Math J.* **139** (2007), 185-202.

[36] A. Rosenberg, The number of irreducible representations of simple rings with no minimal ideals, *Amer. J. Math.* **75** (1953), 523-530.

[37] S. Shelah, Infinite abelian groups, Whitehead problem and some constructions, *Israel J. Math.* **18** (1974), 243-256.

[38] ———, A compactness theorem for singular cardinals, free algebras, Whitehead problem and transversals, *Israel J. Math.* **21** (1975), 319-349.

[39] ———, *Proper Forcing*, Lecture Notes in Mathematics **940** (Springer, 1982).

[40] T. A. Skolem, Some remarks on axiomatized set theory, in *From Frege to Gödel*, J. van Heijenoort, ed. (Harvard, 1967), 290-301.

[41] R. M. Solovay and S. Tennenbaum, Iterated Cohen extensions and Souslin's

problem, *Ann. of Math.* **94** (1971), 201-245.

[42] K. Stein, Analytische Funktionen mehrerer komplexer Veränderlichen zu vorgegebenen Periodizitätsmoduln und das zweite Cousinsche Problem, *Math. Ann.* **123** (1951), 201-222 (German).

[43] S. Tennenbaum, Souslin's problem, *Proc. Nat. Acad. Sci. USA* **59** (1968), 60-63.

[44] S. Todorčević, *Partition Problems in Topology*, Contemporary Mathematics **84** (AMS, 1989).

[45] J. van Mill, An introduction to $\beta\omega$, in *Handbook of Set-Theoretic Topology*, K. Kunen and J. E. Vaughan, eds. (North-Holland, 1984), 503-567.

[46] B. Veličković, Applications of the open coloring axiom, in *Set Theory of the Continuum* (Springer, 1992), 137-154.

[47] ———, OCA and automorphisms of $\mathcal{P}(\omega)/\text{fin}$, *Topology Appl.* **49** (1993), 1-13.

[48] J. von Neumann, Über die Definition durch transfinite Induktion und verwandte Fragen der allgemeinen Mengenlehre, *Math. Ann.* **99** (1928), 373-391 (German).

[49] N. Weaver, Set theory and C*-algebras, *Bull. Symb. Logic* **13** (2007), 1-20.

[50] E. Zermelo, Beweis, daß jede Menge wohlgeordnet werden kann, *Math. Ann.* **59** (1904), 514-516 (German).

[51] ———, Untersuchungen über die Grundlagen der Mengenlehre, *Math. Ann.* **65** (1908), 261-281 (German).

Notation Index

Subject Index